Metals that Shaped Civilization

Metals that Shaped Civilization

The Alloy Secrets Behind Power and Trade

Charles Kensington

Hidden Histories
Media

Staunton, VA 24401

Printed in the United States of America

First Printing, 2025

Published by Hidden Histories Media

Paperback ISBN: 978-1-971137-14-8

Contents

Introduction

Molten metal spills from the crucible like sunlit water, blindingly bright in the darkened room. Sparks leap into the air, tiny stars struggling to survive their brief flight before falling silent on the workshop floor. A metallurgist in a modern laboratory lifts a glowing rod with precision-grip tongs, his protective visor reflecting a light that once came only from charcoal fires in stone hearths. He lowers the metal into a vat, and the liquid hisses violently, steam erupting upward. The sound is primal, ancient, more reaction than technology, more instinct than science.

In that flash of fire and fury, the past wakes up. The same elemental drama played out in the hands of blacksmiths in the deserts of Persia, the forests of Africa, the courts of the Han emperors, and the mountain villages of Japan. It echoed through Bronze Age shipyards, Roman battlefields, and forgotten furnaces buried beneath Saharan sands. The same transformation that happens here, right now under fluorescent lights and digital clocks, lit a fire that has burned with humanity since the first sparks were drawn from stone.

Even today, in an age of satellites and superconductors, the soul of civilization is still forged in heat.

A Question Formed in Fire

How did the earliest metallurgists learn to conquer heat so intense that it reshaped the earth's hardest materials? How did forgotten craftsmen coax steel into bending without breaking, bronze into resisting the sea, and meteoric iron into holding the memory of stars? Why did some technologies flourish into new eras while others vanished in silence?

Metal tells the story of our survival. It enabled agriculture that fed cities. It equipped armies that defended borders and conquered neighbors. It built ships that carried ideas, trade, and ambition across oceans. It raised temples, bridges, canals, palaces, and pyramids. It minted currency and power. It defined ages. Yet the history of metal is not simply progress. It is loss, forgetting, and rediscovery.

Entire civilizations once mastered techniques we still struggle to understand. Damascus steel with its flowing patterns and almost supernatural sharpness. Roman flexible glass rumored to bend like a reed. African iron furnaces capable of temperatures medieval Europe could not match. Korean brass that fueled commerce. Bronze workshops on the Mediterranean seafloor that once linked continents.

Knowledge that required centuries to perfect could disappear in weeks, lost to war, plague, secrecy, or the collapse of trade.

Every slag heap is a library burned. Every broken blade is a question. Every rediscovered ingot is an invitation. This book is about the metals that armed armies, built cities, fueled myths, and then, sometimes, vanished.

The Human Bond with Transformation

Metal demanded a new kind of human. A person unafraid to put their hands near fire hot enough to blind, to blister, to kill. A person who could recognize success by the color of a glow, the ring of a hammer strike, or the scent of ore as it opened. These early metallurgists learned what no one had known before them: earth can change. Nature does not dictate its shape forever.

They saw beauty and danger in molten rivers. They found purpose in chiseling perfection from impurity. They experienced power that had once belonged only to storms and lightning. To command metal was to command destiny.

The earliest blades, formed from stone that fell burning from the sky, were believed to hold divine strength. Bronze alloying gave birth to diplomacy, expansion, and wealth. Steel introduced fear to battlefields that had long felt safe behind bronze shields. Metallurgy reshaped authority and rewrote social order.

Where metal advanced, civilizations advanced. Where metal stagnated, civilizations fell behind. The forge is more than a workplace. It is a crucible where nature and ambition collide.

A Journey Across Continents and Centuries

In these pages, we travel through the deserts of Arabia and Persia, the rain-drenched mines of Cornwall, and the shipwrecked depths of the Mediterranean. We follow artisans who folded blades that Crusaders feared to parry, traders whose cargo of copper and tin steered the fate of Bronze Age empires, priests who carved the heavens into iron daggers buried with kings, and blacksmiths who kept knowledge sacred, sometimes saving it and sometimes damning it to extinction.

The ancient world was more connected than many imagine. Metal moved along routes that predated maps. Ore traced lines invisible but undeniable between Africa and Europe, the Middle East and India, China and the empires beyond. Every blade, buckle, and chisel carried the fingerprints of continents.

Yet metallurgy did not simply spread, it diversified. China's cast iron bore no resemblance to Japan's folded steel. West African bloom furnaces differed vastly from Persian wind-driven towers. Experimentation thrived independently, proving that innovation does not belong to one people, one river valley, one age. The history of metallurgy is the history of humanity discovering its own potential again and again.

What We Lost, What We Seek to Learn Again

Modern scientists break open ancient weapons with the precision of surgeons. Microscopes reveal carbon microstructures that mimic modern composites. Chemical signatures expose trade secrets once carried only by ship captains and caravan guides. Graduate researchers re-create lost alloys in pursuit of performance once achieved by intuition, ritual, and memory. We have modern furnaces that touch the edge of atomic transformation. We have computational metallurgy, robotic assembly lines, and space-bound alloys.

And yet, we still chase the brilliance of vanished masters. Their methods were not written down. Their breakthroughs died with them. Their triumphs and tragedies remain encoded in rust and residue. The earth refuses to give up every answer easily, but it gives up enough to make us hunger for more.

Into the Forge of History

Stand here at the threshold of a story older than writing and more durable than stone. Fire still hums in this world, waiting, glowing, promising. Once we begin, we will uncover metals that should not have existed in their time. We will question how societies that lacked mass communication produced standardized weapons. We will learn why some materials gained mythic status and why even myths contain truth.

We will visit workshops where innovation triumphed, courts where invention was feared, and battlefields where technology changed fate in a single clash.

This is not a story of metallurgy as science. It is metallurgy as revolution. If the foundations of civilization were laid in mud and stone, its walls, weapons, and wonders were raised in metal.

Invitation into the Fire

Walk beside the modern metallurgist once more. Feel the blast of heat as molten steel rolls like thunder into form. Hear the hiss of quenching liquid sealing strength into structure. See the sparks rise, tiny messengers reminding us that every creation is born from risk.

Metal has always been the frontier where imagination meets possibility. It is our most enduring language of innovation.

And even now, as engineers design alloys to build colonies on other planets, humanity feels the same awe the first smith felt when a hammer strike changed the world. We are not so different from those who once smelted copper in clay crucibles beneath desert stars. We are heirs to their courage.

As you turn the page to Chapter One, step fully into the forge. Discover what fire remembers. Discover the secrets civilizations tried to hide. Discover how metal made us and still remakes us today.

Chapter 1 The Sword That Remembered — Damascus Steel and the Blade That Changed Empires

Sparks dance upward as a hammer crashes against glowing metal in the cool of a desert night. The forge breathes fire into the darkness, its heat bending the air and throwing flickering light across the determined face of a master smith. He works beneath a sky filled with stars so numerous they seem to press down upon the earth. A white-hot bar of steel hisses when he quenches it, steam curling upward like a whispered warning. Again, the blade returns to the coals. Again, the hammer shapes it. The rhythmic blows echo like distant drums of war, steady and relentless. As the smith folds the steel upon itself, layer into layer, something hidden begins to take form within. Beneath the blackened scale lies a pattern the eye cannot yet see but the metal already remembers, a flowing design that will one day shimmer like water under sunlight. This is no ordinary blade. It is a weapon destined to strike fear into armies, to define power across continents, and to become a mystery that survives long after empires fade.

The Origins of a Marvel, Wootz Steel in Ancient India

The story of Damascus steel begins far from Damascus itself, in the lush south of the Indian subcontinent. Here, as early as 300 BCE, metallurgists practiced a skill unmatched in much of the ancient world. They crafted wootz steel, an ingot made through a carefully controlled process in sealed crucibles. Instead of heating iron to the point of melting away impurities, these craftsmen combined pure iron with natural carbon sources such as leaves, wood, and plant fibers, allowing carbon to dissolve evenly throughout the iron.

The result was a high-carbon steel that behaved unlike anything known elsewhere. Too much carbon made a blade hard yet brittle. Too little created a weapon too soft to hold an edge. Wootz steel struck the perfect balance, forming within its structure tiny carbides that later produced the famous rippling patterns seen on Damascus blades.

Trade routes carried these valuable ingots far beyond India. Caravans and ships transported them along the Persian Gulf and into Arabia. Indian steel soon became a prized commodity in bustling markets where Middle Eastern craftsmen sought only the finest materials for their workshops. The ingots were small, heavy, and expensive, but inside them lay a hidden potential that would change the future of warfare.

Damascus, Where Knowledge and Flame United

By the first millennium CE, Damascus, one of the ancient world's most influential trade and manufacturing cities, became the center of swordmaking that perfected the transformation of wootz into legendary weapons. Here, Islamic blacksmiths developed forging and heat-treating techniques that preserved and enhanced the unique internal structures of the steel.

Their methods were closely guarded secrets. Temperatures had to be precise. Heating could not be rushed. Folding and hammering required timing that could not be taught by words alone. The smiths quenched blades in mixtures they protected fiercely, combinations lost today because they were whispered from master to student and nowhere written.

When the blades emerged, polished and gleaming, they displayed patterns like flowing water, swirling lines that seemed to move even when the sword was perfectly still. These patterns were not decoration but proof of quality. Inside the steel, carbide bands formed to give the blade a remarkable combination of sharpness and flexibility.

Damascus steel became a cultural treasure. Warriors wielding it felt empowered by something that bordered on the mystical. The blades outlasted their owners. They could bend without breaking. They sliced through helmets and inferior metal as though cutting fruit. Some blades were said to be so sharp that they could slice silk drifting through the air.

There was pride in these weapons, but also fear. A sword that behaved as though it had a will of its own was respected. And every line in its surface was a reminder that a secret dwelled within.

Crusaders Confront the Impossible

Europe first encountered Damascus steel in the medieval Holy Land during the Crusades. Western soldiers carried heavy broadswords designed for crushing armor through weight alone. When these blades struck the lighter, deadlier scimitars of Middle Eastern warriors, the European metal often splintered or bent.

Accounts from the era reveal an awe bordering on terror. Chroniclers described weapons that cut through shields and helmets with chilling ease. The psychological effect was immense. Crusaders believed their foes possessed enchanted metal or hidden knowledge sent from heaven.

Saladin is often symbolically linked to such blades. Though historians cannot confirm whether he wielded a true Damascus sword, the image of the Sultan commanding forces armed with superior steel fueled a mythology that spread throughout Europe and persisted for centuries.

What frightened enemies most was not just the sharpness of the blade, but how it performed under stress. It refused to break. It refused to dull easily. It held an advantage not through brute force, but through superior science long before metals were examined under microscopes.

The Science Inside the Blade

Modern metallurgists have scrutinized surviving Damascus swords using advanced tools. Their findings reveal why these legendary blades surpassed others of their time.

Carbide layers formed in the steel serve as microscopic saw-teeth along the cutting edge, renewing sharpness as the blade wears. The metal is both hard enough to hold an edge and flexible enough to absorb impact, a rare combination made possible by the exact composition of wootz and the forging method used in Damascus.

Even more astonishing, researchers have detected nano-scale structures within authentic Damascus blades. Carbon nanotubes, structures celebrated in today's high-tech materials, appear to have formed during the ancient forging process. The smiths did not know the science behind what they were doing, but they achieved a result modern tools can barely replicate.

And yet, science has not fully solved the riddle. Many attempts to recreate Damascus steel have produced materials similar in appearance but lacking the same mechanical properties. The exact alignment of chemistry, craftsmanship, ore quality, and tradition remains elusive. The sword remembers how it was made. We do not.

Secrets Guarded and Secrets Lost

Knowledge of Damascus steel did not disappear suddenly. It dimmed over time, the flame growing weaker with each generation. Several forces contributed to its decline.

First, changes in Indian ore quality disrupted the composition of wootz ingots. The exceptional characteristics of the finished blades required specific trace elements found only in certain mines that eventually ran dry or were abandoned.

Second, political and economic upheavals disrupted the trade routes that once carried steel from India to Syria. When the flow of wootz slowed, swordsmiths experimented with substitutes. Without the same materials, their techniques became harder to preserve.

Finally, secrecy itself played a role. Masters kept knowledge locked within their workshops. When a blacksmith died without passing on his craft completely, pieces of the method vanished forever.

By the eighteenth century, true Damascus steel had ceased to be produced. Replicas continued, using pattern-welding to mimic the rippling designs, but they lacked the internal structure that made the original material so extraordinary.

An entire technology, one that altered history, simply slipped away.

Legends That Live Where Knowledge Does Not

In the absence of understanding, legends flourished. It was said the blades were forged with sorcery. Tales claimed the steel was quenched in the bodies of slaves so the sword might drink strength from human spirit. Others insisted swordsmiths used powders made from meteorites or rare herbs to enchant the metal.

These myths hold no scientific truth, but they reveal the awe Damascus steel inspired. When engineering outpaces imagination, people fill the space with wonder.

What we know is impressive enough. Damascus steel combined trade, craftsmanship, and innovation into a weapon so advanced it reshaped battles. The closer historians get to the facts, the more remarkable the truth becomes.

A Legacy That Cuts Across Time

Today, Damascus swords rest in museums from Istanbul to London. Curators and scientists handle them with reverence. Collectors and bladesmiths chase the dream of re-creating the ancient secrets. Engineers marvel at the natural elegance of materials crafted without industrial machines.

Its story is not just one of warfare. It is a story of technology shared across continents, of cultures connected through commerce and conflict. It is a reminder that innovation rises and falls with the fortunes of those who wield it.

The rippling patterns on Damascus steel continue to move hearts and spark curiosity. They testify to a forgotten mastery. They ask a question that remains unanswered. How did ancient smiths achieve perfection without fully knowing why it worked?

The blade remembers every fold, every strike, every secret. The world continues to seek what only the sword itself still knows.

Chapter 2
Stone from the Sky — Meteoric Iron and the First Sacred Weapons

A streak of fire slices through the quiet night, bright enough to outshine the stars for a brief and breathtaking moment. People look up, startled, as the heavens seem to tear open. The burning fragment travels with a roar, trailing sparks like a comet determined to reach the earth. When it strikes, the impact shudders across the land. Heat radiates from the crater. Smoke coils into the air. By morning, the fire is gone, but something extraordinary remains. A dense, dark piece of metal rests where no rock should exist. It is heavier than anything else around it, striped with a metallic shimmer that catches the new sunlight. To those who discover it, this is not merely stone. This has fallen from the realm of gods.

Thousands of years later, an Egyptian craftsman sits cross-legged in the dusty light of a royal workshop. Before him lies a sliver of that same strange metal. Copper tools struggle against its stubborn hardness. Every scrape takes effort, every strike a fight. Yet he continues patiently. Pharaohs deserve nothing less than the sky's own gift. What he shapes will be placed beside a king in death and guarded into eternity. Metal from the heavens will protect a divine ruler.

A Heavenly Mystery in the Age Before Iron

Meteoric iron marks humanity's first encounter with iron long before anyone learned how to smelt terrestrial ore. Early societies lived in what modern scholars call the Stone Age, then the Bronze Age, but high-quality iron tools were still a distant dream. Furnaces could not reach the temperatures needed to break chemical bonds inside iron ore.

So when iron arrived pre-formed, already hardened and alloyed, it was a miracle. These fragments did not require mining. They did not come from within the earth. They fell from above, fully prepared to shape into something meaningful.

Most early examples discovered by archaeologists are tiny: beads, small knives, simple ritual objects. That limited size reflects the rarity of the source. Meteorites did not drop daily. Villages did not find harvests of iron scattered along riverbanks. When a meteorite struck, its gift was selective. The metal it granted was limited, precious, and mysterious.

Meteoric iron contains high levels of nickel, often more than ten percent. That unusual composition offered clues even in ancient times. It tarnished differently than other metals. It rang differently beneath a hammer. And most intriguing of all, it seemed unearthly in every sense.

Where did such metal come from? People answered the only way they could: through myth.

Egypt and the Iron of the Sky

Ancient Egyptians looked to the heavens for meaning. The night sky held gods, order, and stories of creation. When iron fell from that sky, it carried sacred power. Egyptians named it "biꜣ-en-pet," meaning iron of heaven or iron from the sky. The name alone reveals how they understood its origin.

Iron beads found in burials as early as 3,200 BCE show the value they placed on this strange metal. These beads were hammered from meteorite fragments millennia before Egyptians mastered iron smelting. To wear iron meant to wear the sky.

The connection only deepened over time. When Tutankhamun was buried in the Valley of the Kings around 1323 BCE, artifacts surrounding him included a dagger made of meteoric iron. The blade was finely polished, dark yet gleaming, fitted with a golden hilt inlaid with gemstones. No ordinary weapon could lie beside a god-king. The dagger symbolized divine protection. It linked Tutankhamun to the celestial world he was believed to join.

Crafting such an object was no simple matter. Copper chisels scraped slowly against the nickel-rich iron. Heat treatments were difficult to control. The metal resisted easy shaping. Every step confirmed that this was a material unlike others.

Yet the difficulty only heightened its significance. A king deserved what the gods sent.

The Arctic's Celestial Tools

Far from desert sands and pyramids, another ancient people shaped stones from the sky into objects essential for life. In the northern reaches of Greenland, Inuit cultures discovered the broken remnants of the massive Cape York meteorite. Unlike cultures that used the metal for symbols of royalty, the Inuit worked meteorite iron into tools of survival.

Harpoons, knives, and scrapers forged from meteoric iron offered a durability that bone and stone could not match. In a frozen environment where food was scarce and every hunt mattered, a sharper edge could mean life itself.

The processing method was simple but ingenious. Heavy stones pounded and fractured the meteorite until workable pieces emerged. These actions connected families not only to the physical earth around them but also to the sky above. Meteoric iron gave them a quiet advantage, a dependable material in a land where nature allowed few mistakes.

Centuries later, when Western explorers arrived, they were astonished at these tools. Many could not understand how an Iron Age community existed without smelting furnaces or mines. They did not yet recognize that these blades were pieces of a cosmic event, fallen long before recorded history.

The largest fragments of Cape York were removed in the late nineteenth century and transported to museums. Even as they became objects of scientific curiosity, they remained evidence that for some cultures, the heavens once provided the means to endure.

Heavenly Iron in the Far East

Early China also left behind hints of meteoric iron's prestige. Archaeologists have found ancient rings, blades, and ornaments whose chemical composition confirms a celestial source. Inscriptions and records speak of heavenly metal, sometimes associated with omens, often with rulership.

In societies that believed rulers served as intermediaries between heaven and earth, possessing a material that visibly bridged both realms offered political power. Wearing meteorite iron was not fashion. It was legitimacy made tangible.

When Fire Learned to Conquer Stone

For thousands of years, meteorites provided humanity's only workable iron. Then everything changed. Around the early first millennium BCE, new smelting methods emerged, likely spread through the Hittites of Anatolia or developed in multiple regions simultaneously.

This was the true birth of the Iron Age. Suddenly, iron did not have to fall from the sky. It could be taken from deep within the earth.

The transition was profound. What had once been a sacred rarity became a resource for farmers, soldiers, and builders. The shift reshaped civilizations. Economies changed. Armies grew larger and more deadly. Tools became stronger, allowing agriculture to flourish in lands where bronze farming tools had failed.

Meteoric iron slowly lost relevance not because its quality diminished, but because its scarcity limited its usefulness in a world now driven by expansion and conflict. The divine metal could not compete with abundant ore that human ingenuity had unlocked.

What had once belonged to gods and kings became common property. Wonder gave way to industry.

The Science Behind the Sky's Signature

Today, modern metallurgy can identify meteoric iron with certainty. One clue is the nickel content, always higher than that of terrestrial iron. Another is found in a set of patterns revealed only through patient polishing and careful etching. These designs, called Widmanstätten patterns, form during the slow cooling of an iron-nickel alloy inside an asteroid, a process that can take millions of years.

No furnace on Earth can reproduce those exact crystalline lines. They are fingerprints of space, proof that the metal slept in darkness before falling into our world.

Scientific tools, including X-ray spectroscopy, magnetic tests, and structural imaging, have confirmed that artifacts found in tombs, caches, and archaeological layers originate beyond Earth. In doing so, researchers have given new meaning to old myths. The ancients were correct about one thing. The metal truly was not of this world.

From Myth to Memory

Meteoric iron sits at a turning point in human history. It represents the first step into a metallurgical future and a reminder of an earlier worldview, when every unexplained event felt sacred. People looked upward and saw mystery. When pieces of that mystery fell into their hands, they responded with reverence, fear, and creativity.

The earliest iron weapons did not swing in great wars. They did not outfit entire armies. Instead, they protected spirits on their journey into the afterlife or supported families who fought the elements to survive another winter.

Every meteorite object created was a marriage of cosmic power and human ambition.

When smelting spread, it brought monumental change. But it also pushed something beautiful into the background. Iron became practical. It lost the aura of the divine. Forging became more common, but the poetry of fallen stars faded from daily life.

A Sacred Beginning

Meteoric iron gave humanity its first lesson in metals: strength can come from elsewhere, and value is shaped not only by scarcity but by meaning. Each culture that touched iron from the heavens formed stories that linked the material to something larger than themselves. Whether in royal tombs of Egypt, family shelters of the Arctic, or palace rituals in ancient China, meteoric iron served the same purpose. It connected humans to the sky.

Later ages would wield iron with unstoppable force. They would build empires with forges that glowed day and night. But before mastery came wonder. Before technology came faith.

The first iron blades were sacred not because they were sharp, but because they were stars that had fallen into human hands.

Chapter 3
Fire That Melts Mountains – Chinese Cast Iron and Imperial Power

A roar like a living creature rises from the furnace. Flames surge upward, lighting the vast foundry hall in rippling gold. Workers with soot-blackened faces hurry beneath the glow, guiding molten metal as it streams out of a smelting hearth and into prepared molds. The air shudders with the pumping of bellows, forcing great breaths of wind into the fire until the heat surpasses anything known beyond these walls. Sparks drift across the scene like swarming fireflies, catching reflections in polished bronze mirrors hung around the hall to magnify the light. In this storm of fire and effort, China shapes power. The Han Empire stands on iron feet and, with each new tool and weapon cast in these blazing workshops, its strength grows.

The heat here is not ordinary. It reaches above one thousand one hundred degrees Celsius, hot enough to melt iron itself so it pours like a river of metal. While other parts of the ancient world still struggled to heat ore enough to hammer iron into shape, China turned the metal into a liquid, allowing it to fill any mold as easily as water. Where others chipped and hammered, China cast. That single difference would change the world.

A Breakthrough of Heat and Ingenuity

Iron is stubborn by nature. It clings tightly to the stone from which it forms. Early blacksmiths in many lands learned to coax iron from ore through bloomery furnaces, laboring to produce small lumps of workable metal. These were then hammered into shape, an exhausting craft that limited both the size and the availability of iron tools.

Yet by the late Zhou dynasty, and especially under the Han, the Chinese achieved something revolutionary. They built taller furnaces with more efficient bellows and preheated air, a technology that forced temperatures far beyond what bloomeries could reach. This innovation allowed them to melt iron completely, changing its nature at the chemical level by saturating it with carbon. Too much carbon made iron brittle and prone to breaking, but it also made the metal melt more easily.

The Chinese learned to control this transformation. They poured molten iron into stone or clay molds, creating shapes that no hammer could easily achieve. Ploughshares, pots, cauldrons, coins, and later architectural beams all emerged from the intense light of the foundry. What had once been a rare material, reserved for the wealthy or for warriors, became the backbone of daily life.

This leap did not come from one discovery or one spark of genius. It was the result of generations. But once it began, the change was swift and undeniable.

Iron in the Fields, Feeding an Empire

An empire cannot survive on prestige. It must eat. Across China's vast river valleys and fertile basins, farmers struggled to tame clay-packed soils and unpredictable floods. Bronze tools, though workable, could not cut deep enough into hardened earth. Wooden ploughs broke easily. Population growth threatened hunger.

Cast iron changed everything. With stronger blades on their ploughs and hoes, farmers turned soil more efficiently, preparing fields for larger harvests. Irrigation tools improved. Permanent iron shovels and picks replaced fragile tools that wore out quickly. Fields expanded, food increased, and families grew.

A state that controls food controls stability. Iron helped feed millions, allowing cities to swell with officials, scholars, artisans, and soldiers. The Han dynasty's prosperity grew from a foundation of iron lines scoring the soil season after season.

Iron for Warriors and Empire

China's enemies feared more than its vast armies. They feared the weapons those armies carried.

Han blacksmiths and foundry masters armed soldiers with cast iron arrowheads produced in enormous quantities. Armor plates shielded warriors from harm, and swords, though not always as flexible as those forged in other cultures, were easier to replace and distribute across far-flung territories.

Casting also allowed experimentation with new shapes and forms. Crossbow triggers made with precise molds granted Chinese infantry a deadly range advantage. Iron spearheads, crafted in standard sizes, made supply chains more reliable. To rule a continent-sized realm, consistency matters.

This industrial approach to warfare marked a turning point. Armies no longer fought with weapons made one by one at a village forge. They fought with standardized arms produced in state-run facilities. Hannibal and Caesar commanded brilliant forces far to the west, but they never marched under the strength of a cast iron arsenal. That advantage belonged to China.

Iron Roads, Iron Bridges, Iron Dreams

Metal flowed not just into weapons but into the infrastructure that bound China's empire together.

Sections of the Great Wall incorporated iron clamps to secure stones in place. Bridges used iron supports to add durability. Waterworks that redirected major rivers depended on cast iron pipes and fittings to manage flow and prevent collapse.

Coinage too benefited from cast iron. Though bronze remained a favored currency, iron coins in some periods allowed the government to stretch resources further. Money traveled more widely. Trade prospered.

It is easy to overlook these objects, as they are not as dramatic as weapons or royal regalia. But they allowed the empire to function. Without stable bridges, food would not reach cities. Without water systems, farms would return to desert. Without iron currency, markets would fail.

Iron built a network of control and communication. The empire became more than a collection of cities. It became a single system.

The State Behind the Furnace

A technological breakthrough can fail if not properly organized. China moved beyond innovation into administration. The Han state placed iron under strict oversight. Regional ironworks were established. Officials recorded quotas and regulated production. Smugglers who attempted to sell iron outside state control risked harsh punishment, for iron was recognized as a strategic resource as important as grain or soldiers.

Certain regions became known for excellence in casting techniques. Government directives ensured that iron flowed first to the army, then to agriculture, and finally to trade. Where others saw metal, Chinese rulers saw national strength.

For a time, this system brought economic stability. But state intervention also introduced difficulties. Too tight a grip could choke innovation. Corruption could warp distribution. The story of Chinese iron is a constant balancing act between centralized control and the needs of the people.

Still, the outcome is clear. Iron transformed the Han dynasty from a powerful kingdom into an imperial giant.

A World Left Behind

Centuries later, Europe would struggle through the Middle Ages still guided by low-temperature ironworking. Craftsmen hammered blooms of iron under strain, relying on techniques that limited both quality and quantity. Cast iron would not become a major force there until the late medieval period and beyond. Industrialization in the eighteenth and nineteenth centuries finally let Europe achieve what China had begun almost two thousand years earlier.

This delay reveals how far ahead China was. Historians studying the spread of technology note that knowledge does not always travel swiftly. Mountains, deserts, and borders of language can block even the most transformative ideas.

While Rome marveled at concrete aqueducts and Greece celebrated its philosophers, China quietly reshaped the meaning of metal itself.

Secrets in the Fire

Though China mastered cast iron before any other civilization, mysteries persist. Some ancient iron objects display remarkable purity and durability, leading modern metallurgists to ask how early Chinese smelters achieved such quality.

Did they refine carbon content more precisely than once believed?

Did they develop advanced molds and alloys now lost?

Were certain ores uniquely suited to the task?

Archaeologists working in preserved ironworks have uncovered scorched ruins filled with clues. Brick-lined furnaces that reached astonishing temperatures. Remnants of complex tuyere systems that directed airflow efficiently. Tools and molds suggesting standardization and high output.

Yet for all we have learned, the precise day-to-day practices of those ancient workers remain largely silent. The process lived in human memory, passed from master to apprentice. When wars, strife, or neglect interrupted that chain, some knowledge flickered out.

The Empire Cast in Metal

Cast iron shaped more than physical tools. It shaped ideas about power. The ability to melt mountains and reshape them into ploughs or weapons gave Chinese rulers a sense of command over nature itself.

An emperor who controlled iron production controlled the means to build, to defend, and to feed. People living under such a system came to see their rulers not only as political leaders but as stewards of elemental forces. Metal was no longer merely a resource. It was a symbol of civilization.

The foundry hall, with its inferno of creativity and its carefully managed workforce, became a microcosm of the empire. Fire, directed by human will, promised a future where strength came not from lineage alone but from knowledge and precision.

The Enduring Legacy of Chinese Iron

Today, archaeologists continue to uncover iron artifacts from the Han period that surprise in both scale and sophistication. Massive cauldrons that once fed soldiers during campaigns. Elegant iron pots that warmed families in winter. Structural reinforcements that held buildings steady against storms.

Cast iron was a quiet revolution. Its rise did not announce itself with single great events or spectacular monuments. Instead, it spread into ordinary life until daily existence depended on it. That is the nature of some of the most profound advances: they become invisible by becoming essential.

Modern engineering acknowledges the brilliance of early Chinese innovators. Their mastery set the stage for later global developments in metal production. And while many countries now produce cast iron in unimaginable quantities, the roots of that power reach back to the roaring furnaces of ancient China.

The empire that melted mountains left behind a legacy visible every time molten metal pours into a mold. The fire that filled those Han foundries still glows in the world around us.

Chapter 4
The Bronze That Ruled the Seas — Uluburun and the Age of Maritime Metal Trade

A quiet world lies beneath the sparkling waters of the Mediterranean. Fish dart between coral formations and ancient timbers. Soft currents stir sand that has settled here for more than three thousand years. There is almost no sound, yet the scene tells a story of journey and disaster. Resting on the seafloor is a ship that once carried treasure across the waves: great copper ingots shaped like ox hides, bronze tools frozen in time, and luxury items meant for powerful rulers. Sunlight shafts down and catches the metallic glimmer of an artifact half buried in sediment. A diver brushes away a thin crust of coral and reveals the past. The Uluburun shipwreck, located off the coast of modern-day Turkey, is a chapter of history written in metal.

A Ship Frozen in a Moment of Crisis

The Uluburun ship sank around 1300 BCE, during the height of the Late Bronze Age. Its cargo represents a world deeply connected through long-distance trade, where essential metals like copper and tin traveled from distant lands to meet the demands of powerful empires. The ship carried enough raw copper to create thousands of bronze weapons and tools. This alone illustrates how central metal had become to economies, armies, and royal courts.

When divers first surveyed the wreck in the 1980s, they discovered more than thirty copper ingots resting in layers, with many more hidden beneath them. These ingots resembled the stretched hide of an ox, a form that made them easier for handlers to lift onto ships and swing over the backs of pack animals. This shape was so iconic that the Phoenicians even used it as a symbol of wealth. Along with the copper were tin ingots, scarcer but critical, for bronze does not exist without tin mixed in.

Bronze revolutionized life in the ancient world. It produced stronger tools, more durable armor, and sharper weapons than stone or copper alone. It became so essential that controlling its supply meant controlling power. The Uluburun ship tells the story of that power in motion, supplies moving toward a destination where political and military futures were being shaped.

Sources of Power, Copper from Cyprus

The copper aboard Uluburun likely came from Cyprus, one of the richest sources in the ancient Mediterranean. The Cypriots mined ore in large quantities and smelted it in workshops that produced the heavy oxhide ingots. Copper left Cyprus on merchant ships and sometimes continued by land caravans, spreading metal into the hands of Minoan traders, Hittite administrators, and Egyptian pharaohs.

Cyprus did not rule an empire in the traditional sense, yet its influence stretched across seas. Everyone needed bronze, and copper was the ingredient that made up the majority of that alloy. To understand the economy of the Late Bronze Age is to understand how Cyprus stood at the center of a metal-based network that helped drive cultural exchange and commercial competition.

The Elusive Partner, Tin from Distant Lands

Tin was the rare ingredient that turned copper into bronze. Without it, copper dulled too easily and bent under stress. Historians still debate exactly where most Bronze Age tin originated. Some believe the Uluburun tin may have come from Central Asia or Anatolia. Others point to even farther regions such as Afghanistan or the area now known as Cornwall in Britain.

This uncertainty highlights a reality. Bronze Age trade traveled astonishing distances. Metal moved through hands and across borders that no single civilization controlled fully. Rulers had to negotiate, trade, and sometimes fight to secure steady tin supplies. One weak shipment, one lost deal, could weaken a kingdom's weapons or tools in the generation that followed.

Thus, the balance of Bronze Age power rested not only on armies but on the success of trade.

Bronze in the Halls of Power

The Uluburun ship carried more than metals. Ivory tusks, glass beads, gold jewelry, and exotic tree resin suggest that wealthy clients waited eagerly for its cargo. Among the most intriguing finds was a gold scarab bearing the name of Nefertiti, the queen of Egypt. While not proof that this exact shipment was intended for the pharaoh's household, it demonstrates the kind of elite network that depended on maritime trade.

Minoans on Crete, Mycenaean Greeks on mainland Greece, and Egyptians in the Nile Valley all sought status goods like carved ivory and colorful faience. At the same time, they needed bronze for military strength and agricultural progress. Courtly gift exchanges relied on these desirable metals, and kings used them to reinforce alliances.

A Mycenaean palace bristled with bronze-tipped spears and helmets. Egyptian charioteers went to battle with bronze fittings and arms. Without the constant flow of metals, royal courts could not reward loyalty or prepare for war. The metal trade was not just commercial. It was political.

Mariners of a Connected World

The crew of the Uluburun ship were part of a growing class of merchants who linked kingdom to kingdom. Sailors navigated using stars, memory, and familiar coastlines. They faced storms, pirates, and unpredictable political alliances. Yet they traveled anyway, for the rewards were immense.

Unlike many ancient societies where local production limited what people could access, Bronze Age maritime traders brought goods that crossed cultural lines. Alongside metals, the wreck yielded logs of ebony wood, ostrich eggshell vessels, and thousands of beads crafted in distant workshops.

These shipments carried ideas as well as objects. Religious symbols, artistic styles, and technological knowledge spread alongside ingots of metal.

The oceans were not barriers in the Bronze Age. They were highways.

Statecraft Forged in Bronze

Bronze technology supported not only armies and agriculture but also bureaucracy. Tablets of clay recorded shipments and legal agreements, often with bronze styluses. Bronze keys locked storerooms containing taxes in kind, grain, oil, or livestock. Tools made from bronze helped construct temples and fortresses where rulers claimed divine authority.

Diplomats exchanged gifts of bronze weapons and luxury metalwork to secure peace or demonstrate dominance. The famous Amarna Letters, preserved from Egypt, include negotiations involving metal shipments, requests for materials, and complaints when deliveries failed to arrive.

Metal was not merely raw material. It was leverage.

Cracks in an Interconnected World

For several centuries, the Bronze Age trade network thrived. Then around 1200 BCE, connections began to fracture. Archaeologists link this time to the collapse of multiple kingdoms and city-states across the eastern Mediterranean.

Natural disasters, invasions by the mysterious Sea Peoples, and internal political struggles all pressured the system. But economic breakdown may have been the most devastating blow. If copper could not reach Cyprus's smelting workshops, the chain failed at the source. If tin caravans were disrupted, bronze production fell silent.

Suddenly, kingdoms faced shortages of the metal that had supported their power. Armies weakened. Construction stalled. Trade shrank. A collapse in one region rippled outward, creating a domino effect that ended the Bronze Age's golden connections.

The Uluburun ship may have sunk in this fragile moment. Its cargo, instead of strengthening a kingdom, was lost to the deep.

Rediscovery and Revelation

Thousands of years later, archaeologists brought the Uluburun story back to the surface. More than twenty thousand artifacts have been conserved and studied. Each piece reveals layers of context: where it came from, who may have handled it, and what it was intended to do.

Analysis of the copper shows isotopic signatures matching Cypriot mines. The tin matches ores from far eastern sources. The collection includes personal items, a small wooden writing board, a bronze dagger, and weights used for trade, riveting evidence of the human lives behind the commerce.

The shipwreck changed historians' understanding of the Bronze Age. Instead of viewing ancient societies as isolated, Uluburun demonstrates an international economy long before the concept existed. It proves that the sea served as the central artery of political and economic life.

The Sea's Silent Storyteller

The Uluburun ship shows a world that believed the sea could unite. Across cultural borders, goods and ideas moved freely from Egypt to Anatolia, from Cyprus to the Aegean.

Yet the same sea that made wealth possible also concealed disaster. When winds turned or enemies attacked, ships sank and knowledge went with them. Empires needed constant vigilance against nature and rivals alike.

What we recover from shipwrecks is often incomplete, but enough remains to remind the modern world of a truth that Bronze Age sailors lived daily. Human progress depends on cooperation and the ability to maintain connections across distance. When those connections fail, civilizations stumble.

The Alloy That Built Civilizations

Bronze was more than a metal. It transformed societies by enabling more intensive farming, more powerful militaries, and more complex economies. It allowed rulers to concentrate power and expand influence. It demanded organization, negotiation, and shared purpose. And it came from the sea.

The Uluburun shipwreck rests today as a museum piece and a message. Inside its aged timbers lies a moment when the Bronze Age stood near its height, unaware of how fragile it was. The copper and tin aboard were destined to shape tools and weapons that would build and defend kingdoms. Instead, the sea claimed them, preserving their story for us.

Bronze once ruled the Mediterranean's waters and the great civilizations around them. Even in silence beneath the waves, it continues to speak, telling of merchants who dared the sea, of kings who demanded metal, and of an age when the strength of a civilization could be measured by the shine of its bronze.

Chapter 5
The Emperor's Hidden Metal — Korean Hwanggeum and the Power of Brass

Inside the blazing heat of a Korean forge, glowing liquid metal flows like molten sunlight. Sparks whirl upward as artisans carefully tilt crucibles, guiding the shimmering alloy into waiting molds. Newly cast coins clink as they cool, their surfaces gleaming with a gold-like shine that catches every flicker of firelight. The workers look on with pride. Visitors from distant shores stare in astonishment. This is no ordinary metal. It resists corrosion. It holds its luster. It shines with a warmth that evokes wealth and power. It is hwanggeum, a brass alloy so advanced and so coveted that it helped define Korea's role in the ancient East Asian world.

Foreign traders arriving at Korean ports rubbed their fingers across the bright currency, inspecting its quality and heft. What they found surprised them.

These were coins that did not rust in humid air or crumble when exposed to sea salt. Hwanggeum, literally "yellow metal," combined beauty with practicality in a way that other metals of the region struggled to match. Behind each coin lay a secret fire and a metallurgical process that challenged even the most skilled craftspeople beyond Korea's borders.

The Rise of a Golden Alloy

To understand hwanggeum's importance, one must look back to the early Korean kingdoms such as Silla, Baekje, and Goguryeo. In these formative centuries, metalwork was not merely a craft. It was a political necessity. Leaders needed tools to build their cities, weapons to defend their borders, and currency to manage growing trade. Bronze, an alloy of copper and tin, had already shaped Korean culture, most notably in ritual bells and elegant mirrors. But another alloy was rising, one that looked like gold while requiring none of the king's fortune.

Brass, the alloy of copper and zinc, formed the foundation of hwanggeum. Copper was common, but zinc posed challenges. Unlike tin, which melts easily and combines readily with copper, zinc evaporates when heated. To produce a strong, high-zinc brass, Korean artisans needed to understand distillation and timing. They learned to heat zinc ore in separate vessels, allowing the vaporized zinc to travel into molten copper at precisely the right moment. Too slow or too hot, and the zinc vanished into the air. Too cool, and the metals refused to bond.

The mastery of this delicate process arrived earlier in Korea than in many other regions of the world. Where others struggled to create consistent brass, Korea took pride in producing a material that surpassed old bronze in both appearance and performance. Hwanggeum resisted corrosion from coastal environments. It produced clearer tones in bells and horns. It offered a golden sheen attractive in both ceremony and commerce.

A king did not need a treasury full of gold to project authority when he could mint coins and adorn palaces in this brilliant metal.

Coins That Carried Power

The Silla kingdom, which eventually unified the Korean peninsula in the late first millennium CE, embraced hwanggeum as a tool of governance. Coins standardized in shape and inscription flowed across emerging markets. To ordinary farmers, brass currency meant stability. To merchants, its durability meant wealth could survive a journey from inland farms to seaside ports.

Currency holds identity. When a buyer handled a coin minted by Silla, they acknowledged the authority of the crown. The weight, shine, and resilience of hwanggeum reinforced that authority. With a stable and recognizable form of money, trade within Korea strengthened. As demand for these well-made coins spread to foreign lands, Silla positioned itself as a powerful player in regional commerce.

The golden aesthetic appealed to human desire for luxury, yet brass remained far more attainable than true gold. Silla rulers turned this advantage into a political message: splendor could be widespread, not restricted to elites. Soldiers carried brass fittings on their armor. Officials used brass seals and writing instruments. Homes adopted brass bowls that would not tarnish like common metals. Hwanggeum served both palace and village, an early example of a metal shaping national character.

The Secret of Zinc Vapor

The existence of brass in other parts of the ancient world cannot be ignored. Merchants in the Mediterranean had known forms of brass since the late Bronze Age, and the Middle East explored similar alloys. But Korea advanced the craft beyond many rivals through precision and scale.

Zinc does not announce itself in the smelting process. It disappears into smoke unless captured by skilled control. Korean metalworkers developed methods to channel zinc vapor into copper using sealed crucibles and controlled airflow. They experimented with clay-lined furnaces that minimized loss. They learned to judge heat by flame color and metal sound, cues born from experience rather than written records.

Such knowledge rarely traveled far from the forge. Master artisans protected the techniques fiercely, sharing them only with trusted apprentices. The secrecy preserved Korea's advantage and ensured that foreign merchants valued hwanggeum as a unique commodity.

China, with its immense industrial might, invested heavily in bronze and iron yet struggled for centuries to produce brass of equal consistency. Japan admired hwanggeum items and sought to adapt the techniques, but Korean expertise often remained a step ahead.

A smelting secret can be as powerful as an army, and Korea guarded this one well.

Brass Beyond Borders

Trade routes across East Asia did not flow in a single direction. Korean ships carried goods to Chinese ports, and envoys from Japan traveled westward to study culture and craftsmanship. Hwanggeum traveled these lanes with quiet influence. The golden shine of Korean brass intrigued foreign courts. Brass ornaments adorned ceremonial dress. At times, Korean-crafted items were presented as diplomatic gifts, reminders that Korea was not merely a recipient of culture from giants like China but a contributor to shared innovation.

Wherever Korean brass arrived, it represented a sophisticated metallurgical economy. Foreign traders recognized the value not only in the object but in the knowledge behind it. This admiration helped Korea forge alliances and build commercial networks that spanned the Yellow Sea.

One could argue that hwanggeum was a quiet ambassador, building Korea's reputation without the need for boasts or conquest.

Weapons with Golden Strength

Brass may not be suited for swords or arrowheads, as steel surpasses it in lethal hardness. Yet weapons are more than blades. A soldier's survival depends on buckles that hold armor tight, fittings that secure shields, and trumpet signals that relay commands. Brass excelled in these roles. It resisted the elements on campaigns. It allowed artisans to shape decorative accents that symbolized unit pride and royal favor.

In times of ceremony, hwanggeum shone as brightly as victory. In times of conflict, it stood firm while lesser metals fell to rust.

Military power does not come only from the cutting edge. It emerges from every part of an army that must be coordinated, equipped, and united. Hwanggeum strengthened that structure, serving as support behind the steel, as confidence behind the blade.

Bronze, Brass, and Identity

Comparing Korean brass to earlier bronze reveals a fascinating transformation. Bronze carried spiritual weight, often used in bells and ritual objects that honored ancestors and gods. But hwanggeum reflected everyday life. It represented a shift from the ceremonial to the practical, from weapons of the elite to goods for the common people.

Gold symbolized divine favor. Iron symbolized military power.

Brass, hwanggeum, symbolized connection. It linked village markets to palace halls. It linked Korea to its neighbors. It linked wealth to work. It looked precious without requiring exhaustion of gold reserves. It helped unify a growing society under shared currency and shared prosperity.

A Knowledge Slow to Spread

Centuries passed before Europe mastered brass in the same controlled way Koreans had long practiced. In Europe, brass often resulted from accidental contact between copper and zinc-bearing ores, producing inconsistent results. Not until advances during the medieval period did European brass achieve widespread and reliable quality.

This delay underscores the significance of Korea's achievement. Metallurgy requires not only resources but scientific insight and collaborative skill. Korean artisans embraced experimentation early. They pushed furnaces hotter than was common. They observed how vapor behaved. Their reward was a metal both useful and elegant.

When global trade eventually merged East and West, the reputation of East Asian brass traveled across oceans. The techniques honed in Korea helped establish the region as a center of skilled metallurgical innovation.

A Golden Legacy

Modern Korea remembers hwanggeum not merely as a currency metal or an alloy of convenience. It represents ingenuity born from necessity and ambition. Museums showcase ancient coins, bowls, and fittings still gleaming centuries after they were cast. Metallurgists marvel at how uniform zinc distribution was achieved without modern gauges or chemistry. Historians trace how brass supported political unification and economic resilience.

Hwanggeum tells the story of a nation that chose material innovation as a tool of growth. It helped Korea rise into a position of influence in East Asia, not through conquest but through craftsmanship.

Every gleaming coin minted in those early foundries carried with it the message that Korea understood the power of metals not only in forging weapons but in forging identity. When foreign traders lifted one from a merchant's hand, they saw beauty and durability, but also authority, stability, and pride.

The emperor's hidden metal was more than golden brass. It was a symbol of a people shaping their future with fire and imagination, turning the secrets of the forge into the prosperity of a nation.

Chapter 6
The City of Copper Walls – The Indus Valley's Metallurgical Might

Heat thickens the air in a cramped workshop beneath a low, smoke-darkened roof. A fire crackles inside a clay furnace, fed by bellows that rise and fall in a steady rhythm. Sparks leap upward, briefly illuminating the determined faces of smiths who bend glowing metal into shape. Copper melts into a bright, malleable orange, ready for transformation. One artisan pours it carefully into a small mold. Another hammers a cooling strip into a blade that will soon cut grain or carve wood. Outside, a lively market buzzes with traders examining bronze ornaments polished to a warm, reflective gleam. Children chase each other between stalls where bracelets, fishhooks, and sturdy copper tools glint in the sun. This is not a simple village. This is Harappa in the third millennium BCE, a city built on metal.

The Indus Valley civilization, flourishing between 2600 and 1900 BCE, stands as one of humanity's earliest and most impressive urban cultures. Its people constructed vast cities with standardized bricks, complex drainage systems, orderly streets, and workshops filled with skilled craftspeople. Metal powered much of this innovation. Copper and bronze were not luxuries confined to temples or palaces. They were everyday materials that shaped work, trade, and identity.

Masters of Fire and Metal

The people of Harappa and nearby Mohenjo-Daro were among the most capable metallurgists of their age. They smelted copper ore in specially designed furnaces lined with heat-resistant clay. Workers knew exactly when to add air to raise the temperature and how to refine molten metal to remove impurities. Their furnaces reached impressive heat, proving they understood the demanding requirements of metal production.

The resulting copper tools served farmers, carpenters, hunters, and traders alike. Axes chopped timber used in construction. Saws smoothed wooden beams that supported homes. Bronze spearheads and arrow tips gave hunters a deadly advantage when pursuing game. The reliability of these tools supported every major industry of the civilization, from agriculture to architecture.

Their craftspeople reached even higher levels of artistry through techniques like lost-wax casting. This method involved shaping wax into an object, coating it with clay, and heating the mold so the wax melted away, leaving a hollow space ready for molten metal. The final product displayed intricate forms and fine details, often too complex to achieve through hammering alone.

One extraordinary example is the famous bronze statue of the "Dancing Girl" found in Mohenjo-Daro. A small, elegant figurine with a confident stance and jewelry hinting at fashion and status, it demonstrates not only technical mastery but also an appreciation for creativity and human expression in metal.

The Bronze Economy

Bronze, an alloy of copper and tin, expanded the range of tools and goods the Indus Valley could produce. Bronze blades held sharper edges than pure copper. Bronze chisels lasted longer under the strikes of a hammer. Bronze ornaments held their beauty against weather and wear.

To obtain tin for making bronze, the Indus Valley relied on trade networks that reached great distances. Archaeologists believe tin may have traveled from regions such as Afghanistan or Central Asia through established caravan routes. Copper flowed in from mines in Baluchistan, Rajasthan, and Oman.

Resource control became a source of economic stability. Communities that maintained reliable access to ore could support their craft industries, which in turn fueled domestic prosperity and international trade. Metalwork helped create a cycle of success: the more tools workers produced, the more efficiently they could build, harvest, and trade.

Trade That Linked Worlds Together

Evidence from shipwrecks, coastal settlements, and foreign records suggests the Indus Valley civilization was not isolated. Ships sailed across the Arabian Sea carrying copper, precious stones, and manufactured goods. Indus seals, stone stamps carved with animals and symbols, were found in Mesopotamia, thousands of miles away. Copper goods and finished bronze objects traveled this route, demonstrating the region's position at a crossroads of commerce.

Overland trade brought metals into the heart of the Indus cities. In return, skilled artisans exported finished goods admired by distant cultures. Luxurious ornaments, standardized tools, and finely crafted objects marked the Indus people as both producers and consumers within a thriving network.

Copper ingots and bronze blades were not simply commodities. They were messengers that carried the story of a prosperous civilization beyond its borders.

Metal in Daily Life and Urban Innovation

The Indus cities were among the most meticulously planned environments of the ancient world. Streets ran in organized grids. Houses used standardized bricks in precise dimensions. Drains flowed beneath streets to collect wastewater and keep communities clean.

Metal tools made much of this possible. Without durable chisels, standardized bricks were harder to shape. Without sharp axes and saws, wood for roofs and frames required more time and effort. Without strong nails and fittings, large-scale architecture would strain against its weight.

In Harappa's markets, metal objects shaped daily routines. Workers relied on bronze items to process grain and maintain livestock. Fishermen with bronze hooks brought home abundant catches. Merchants used scale weights made of carefully measured stone and metal that ensured honest transactions.

These artifacts show a society that valued fairness, organization, and participation in a shared system of goods and services. Metal supported not only commerce but also trust.

Bronze as Status and Style

Metal also became a marker of beauty and cultural expression. Bronze jewelry adorned men and women alike. Copper mirrors reflected faces in the glow of oil lamps. Metalworkers engraved intricate patterns onto blades and bowls.

It is easy to imagine pride on the face of someone wearing a new bracelet or presenting a gleaming ornament during a festival. These adornments were not merely decorative. They communicated wealth, identity, and belonging.

Metal strengthened social bonds through symbols and gifts that shaped community life.

Mysteries Locked in Silence

For all the material evidence recovered from the Indus Valley, much remains unknown. Unlike Egypt or Mesopotamia, where written language reveals beliefs and history, the script of the Indus remains undeciphered. Clay tablets bearing icons and signs sit silent in museums. Scholars study them tirelessly, yet a full translation has never been achieved.

This leaves major questions unanswered.

Who controlled the metal workshops

Were smiths honored specialists or state employees

Did rulers command trade, or did merchants coordinate independently

The cities are gone. Their rulers unnamed. Their stories left to be pieced together through shards and tools.

Yet the metal tells us what the words cannot: the civilization was advanced, organized, and deeply connected to the wider ancient world.

A Civilization Forged, Then Forgotten

Around 1900 BCE, the Indus Valley civilization began to decline. Climate change, shifting rivers, and economic instability likely played roles. Trade slowed. Cities gradually emptied. Workshops fell silent.

Even after their society faded, the influence of Indus metallurgy persisted. Neighboring cultures inherited techniques. Future civilizations admired and copied the forms their metalworkers perfected. Their innovation laid foundations for technologies that shaped Asia for millennia.

The Legacy Beneath the Earth

Today, archaeological digs continue to reveal the Indus Valley's metallurgical might. Copper tools remain sturdy. Bronze ornaments still carry their sheen. Casting molds and crucibles recovered from workshop floors speak of a tradition driven by precision and pride.

These discoveries remind us that progress does not always leave written explanations behind. Sometimes it leaves melted metal, hammered edges, and the quiet proof of hardworking hands.

The Indus Valley civilization built cities whose strength ran deeper than bricks and streets. Its true walls—those that protected life and supported growth—were the copper tools in workshops and the bronze goods in marketplaces.

Fire, metal, and ingenuity combined to forge not only the tools of civilization but the civilization itself. The world of Harappa sparkled with bronze and shone with copper, a reminder that humanity's earliest great cities were built not only of stone but of metal that fueled imagination and industry.

Chapter 7
Glass That Would Not Break — Roman Vitrum Flexile and the Secrets That Died with an Emperor

Candlelight shivers across an imperial table, making the polished surfaces glow like a shallow pool at dusk. In the hush of the audience chamber, an artisan steps forward with a cup that seems to hold its own pale moon. He bows, then presses his thumbs gently into the vessel's rim. Instead of the expected crack and fatal starburst, the cup bows. A supple arc ripples through it, as though the glass has turned to liquid for an instant and then remembered to be solid again. Courtiers lean in. Armor creaks. Someone swallows. The cup straightens. The room does not. The air has changed, charged with the feeling that some boundary has been crossed. It is said the emperor Tiberius watches without a smile.

The Tale in the Sources

The story of Roman flexible glass, the famed vitrum flexile, comes to us as a rumor wrapped in literature, often retold at dinner tables and in scholars' studies. The version most often cited involves an artisan who demonstrates a bendable glass vessel before Tiberius. After denting the cup, he repairs it with a small hammer, showing that this glass does not shatter like ordinary ware. The emperor, so the tale goes, demands to know whether anyone else knows the secret. Learning that the formula resides with the artisan alone, he orders the man executed and the workshop destroyed, fearing that such a marvel would make precious metals lose their value and disrupt the imperial economy.

Ancient writers transmit pieces of the tale with different edges. Petronius, the courtier and satirist, evokes a world of extravagant banquets and suspect wonders where such a story makes moral sense, even if he never delivers a laboratory report. Pliny the Elder, the encyclopedist of the first century, circles rumors of malleable glass with a natural philosopher's skepticism. Others in late antiquity repeat the anecdote in forms that shift with the teller's meaning. None provides a specimen, a recipe, or a workshop inventory. What survives is the idea, not the artifact.

That absence does not prove falsehood, but it sharpens the historian's task. The tale speaks in the language of court politics and economic anxiety as much as in the language of materials science. To read it well is to read it on both levels at once.

The State of Roman Glassmaking

Rome in the first centuries BCE and CE was a superpower of glass, even without miracles. The breakthrough of glassblowing, likely emerging in the Syro-Palestinian region shortly before 50 BCE, transformed production. Instead of laborious casting or core forming, craftsmen blew molten glass into bubbles and bodies with a speed and freedom that astonished markets. Mold-blowing followed, allowing workshops to stamp patterns, ribs, and inscriptions with repeatable precision. Natron from Egyptian lakes served as the flux that lowered silica's melting point. Lime stabilized the soda-lime-silica blend. Manganese bleached away green tints to give a prized colorless look that imitated rock crystal.

In such furnaces, artisans could do things earlier ages could not: roll out thin panes for windows, shape stacked-thread handles, pull canes of colored glass into mosaic patterns, and produce tableware for patrons from the Aventine to Antioch. Glass did not rival gold for status, but it democratized luster. It could be transparent, tinted, marbled, or layered. It could be blown into a bottle for perfumed oil or a cup for an everyday tavern. Above all, it was brittle. When dropped or struck, it broke with a sound everyone knew.

Against this backdrop, a cup that bent rather than shattered would have felt like an intrusion from another physics. The marvel lands most effectively in a world that already knows how glass behaves.

Why Flexibility Mattered

The artisan's demonstration is dramatic for a simple reason: glass fails predictably. It lacks the dislocation mechanisms that allow metals to yield. Press too hard on a rim and microcracks, invisible at first, race across a surface until a flaw becomes a fracture. To watch a glass cup dent and rebound violates everyday experience. If such a thing existed, it would promise resilience in transport, safety in use, and a new hierarchy of materials. Imagine drinking vessels that survive a fall, window panes that flex in wind, storage jars that shrug off knocks from a cart wheel. In a culture that moved goods by ship and wagon over long distances, survivability translated into fewer losses and lower costs.

But the tale's logic goes further. If glass could perform like metal, then metal's exclusive prestige would weaken. For a court where silver plate signaled rank, a flexible glass cup threatened more than an industry. It threatened symbolism. It blurred boundaries that Rome depended on.

What Could Vitrum Flexile Have Been

To separate myth from plausible science, one must ask what material could create the behavior described while remaining recognizable as glass.

One candidate is metallic glass: an alloy cooled so quickly that its atoms cannot form a crystalline lattice, leaving an amorphous structure with unusual strength and elastic behavior. Metallic glasses today can bend slightly and spring back. Yet their production requires cooling rates and compositional control far beyond ancient capabilities. Moreover, metallic glass looks like metal, not like a translucent goblet glowing in candlelight.

Another possibility is a laminate: a thin glass shell bonded to a ductile metal substrate. If the glass adhered well, the composite might take a dent and recover. But the emperor would have recognized metal at once. The wonder must read unmistakably as glass.

Tempered glass does not help. It is strong but fails suddenly. A tempered goblet, if nicked, is more likely to shatter into pebbles than to bow and right itself.

Ionic-exchange strengthening, the basis for modern phone screens, was far beyond ancient chemistry. Even if a local brine produced minor strengthening by chance, it would not create the dramatic ductility described.

A more plausible scenario involves thin walls and theatrical demonstration. Blown glass with very thin, rounded surfaces can flex minutely under careful thumb pressure. A trained artisan could demonstrate that tiny "give" in a dim palace hall and then tap the vessel to produce a sound that suggests repair. The spectacle becomes the marvel.

Another possibility is that the object was not silica glass at all, but a transparent mineral like rock crystal. Crystal handles pressure differently and can produce effects that appear uncanny when compared to everyday glass.

None of these options proves the story. All show that "flexible glass" could be a mixture of science, skill, and interpretation.

The Court's Reaction and the Economy of Prestige

Tiberius valued fiscal stability. Precious metals underpinned Rome's economy. Silver and gold signaled rank, served as coinage, and decorated elite tables. A material that behaved like metal while appearing as glass threatened the symbolic ladder that defined Roman hierarchy. If flexible glass became fashionable, silver plate might seem less essential. If the middle class could afford such wonders, visual boundaries between classes could blur.

There was also the danger of monopoly. If only one artisan held the secret, he controlled the supply. An emperor accustomed to mastery over resources might see that as intolerable.

The tale, even if exaggerated, reflects real concerns about luxury, scarcity, and control.

Suppression as a Theme in Roman Technology

Rome did not oppose innovation wholesale. It built aqueducts that carried mountains of water and poured concrete that defied time. But Rome also regulated guilds, guarded monopolies, and resisted innovations that threatened entrenched interests.

Glass itself had once been a luxury. When glassblowing spread it widely, elites pivoted to other markers of status. Innovations arrived into a world constantly negotiating value.

Vitrum flexile, therefore, becomes a parable about how societies accept or resist change. A safer glass for storage jars or windows might have been welcomed. A dinner cup that challenged silver's prestige risked condemnation.

What the Sources Tell Us, and What They Cannot

Petronius uses the story to mock excess. Pliny approaches it cautiously, treating it as rumor. Later writers repeat it because it feels emblematic, not because they can verify it.

The story survives because it expresses Roman anxieties about wealth, hierarchy, and the power of craft. It reminds us that technologies rise or fall not by engineering alone but by the societies that judge them.

The Plausible Kernel

Strip the fable to its core and a plausible scenario remains. A glassworker trained in advanced annealing techniques blows a vessel with thin, even walls. He anneals it carefully. It can flex slightly without cracking. In a carefully managed demonstration, he presses the rim and produces a small deformation that reads as a dramatic bend. When he taps the dent out, the vessel resettles. The court gasps. The emperor feels both wonder and unease.

From there, rumor takes over. Execution, suppression, sole inventor—all hallmarks of stories told to explain the absence of evidence. The tale becomes a vessel for cultural truth rather than scientific fact.

Consequences of Innovation Suppression

Even if the event never happened, the story illustrates the cost of discouraging innovation. It teaches artisans that power can be dangerous. It warns that novelty must align with elite expectations to survive. It suggests that certain advances, especially those that challenge status symbols, struggle for adoption.

Rome advanced regardless. Glassmaking improved, markets expanded, and homes brightened with translucent panes. But if techniques existed that could have made glass safer or more resilient, a culture wary of disruption may have slowed their spread.

The Afterlife of a Rumor

The rumor endured because each age found meaning in it. Renaissance scholars saw jealous patrons. Enlightenment thinkers read fears of monopoly. Modern readers see a parable about markets resisting disruptive technology.

Each retelling bends the cup for its own purposes. Yet none breaks it. That is why the story lasts.

What Survives in the Furnace Light

Imagine the artisan leaving the palace by torchlight, whether triumphant or condemned. He returns to a workshop where furnaces breathe like animals and molten glass flowers into bowls with a twist of the wrist. Craft continues, regardless of politics. That is the one certainty.

Roman glassmaking needed no miracles to transform daily life. Transparent vessels changed commerce, medicine, and domestic comfort. The rumor of flexible glass stands at the edge of this record, a thought experiment placed in a marble hall. Whether the cup bent as described matters less than the anxieties the story reveals.

A Measured Verdict

No specimen of flexible Roman glass survives. No recipe, no technical description, no independent confirmation. Yet there are texts shaped by a first-century world attuned to the dangers and promises of innovation.

The plausible science suggests a trick performed by a skilled artisan. The plausible politics suggests a ruler wary of anything that threatened symbolic order. Together, they form a story less about a cup and more about the conditions in which technologies live or die.

In the end, the artisan's hands teach the longest lesson. Materials do not shape history by themselves. They do so when craftspeople are allowed to experiment, when courts and markets reward performance over pedigree, and when wonder is treated as the beginning of discovery rather than a threat to authority.

Chapter 8
Iron for the Gods – The Hittites and the Weapon That Ended the Bronze Age

The sun beats mercilessly on the plains of Anatolia. Soldiers struggle in the churning dust, sweat stinging their eyes, lungs filled with heat instead of air. Bronze shields lock together. Bronze blades spark wildly as they crash against enemy lines. But in one section of the battlefield, something different happens. A sword slices cleanly through a bronze edge as if it were soft wood. The chipped remains of a defender's blade clatter to the dirt. Panic spreads. Opponents falter as warriors armed with darker, sharper weapons push forward. In their hands are tools of a new era, guns before gunpowder, secrets hammered in fire. The black iron of the Hittites glints like lightning in the chaos of war.

The Metal That Changed Power

For centuries, the civilizations of the Near East built their world on bronze. Kings hoarded copper and tin. Trade routes stretched across deserts and seas to feed a metal economy that armed armies and shaped tools. Bronze meant strength. Bronze meant civilization. Bronze meant control.

Then iron entered the story. Iron ore was abundant, scattered in mountains and buried in fields. Unlike tin, it did not require long-distance trade with unpredictable partners. But iron refused to give up its strength easily. Its melting point was far higher than that of copper. Early smelting methods produced spongy blooms of iron that required extensive hammering to drive out slag. Most early societies saw this stubborn metal as inferior to bronze, too difficult to produce well, too brittle if mishandled.

The Hittites, however, were not deterred.

Centered in Anatolia, in what is now Turkey, the Hittite Empire rose to prominence in the second millennium BCE. These were people who combined military ambition with an appetite for innovation. They looked at iron's challenges not as barriers, but as mysteries worth solving. Over generations, they mastered the furnace. They learned how to heat iron ore with charcoal, increase airflow, and refine the bloom until it hardened into a metal stronger and sharper than bronze when treated correctly. They did more than use iron. They learned to transform it.

Sacred Metal, Sacred Power

To the Hittites, iron was more than a material. It was divine. Inscriptions and ritual texts suggest that iron was linked to gods of storm and war. Early iron objects from Hittite lands often appear not as military gear but as ceremonial offerings. Daggers, tools, and axes carried symbolic meaning, gifts meant to honor deities or confirm royal favor.

This sacred connection fostered secrecy. It was not enough to forge iron. One guarded its secrets as one would guard holy knowledge. Artisans trained under the protection of noble houses. Metallurgical techniques became state assets, hidden from traders and rivals. When the sacred became strategic, the Hittites held the advantage.

Forging Dominance in War

The Hittite Empire grew bold. With iron-tipped spears and iron swords, their troops faced neighbors outfitted primarily with bronze. Iron weapons, once hammered and quenched, retained edges that shattered bronze upon impact. Battlefield victories reinforced the value of this new metal.

One defining moment came in 1274 BCE at the Battle of Kadesh between the forces of Hittite king Muwatalli II and the Egyptians under Ramses II. Although the battle ended in a stalemate, both sides declared victory for the sake of honor. What mattered more was the political outcome that followed. The first recorded peace treaty in history was signed between the two powers, and within its diplomatic exchanges lay references to iron.

As Egypt admired Hittite iron, the metal gained prestige equal to gold. Royal letters show that iron-related negotiations occurred between courts, indicating its scarcity and immense value. Some records list iron alongside precious metals, revealing that even small shipments carried major weight.

The Hittites did not mass-produce iron weapons. Output remained limited. But in elite units, iron blades proved decisive. A handful of iron swords could turn the tide against lesser bronze armies.

Power in the ancient world often flowed from monopoly. The Hittites held one on iron.

The Price of Secrecy

While iron ascended, bronze still dominated many aspects of life. It remained easier to work and could be produced in bulk so long as trade routes remained stable. But those routes were fragile.

When disruptions rippled across the Near East around 1200 BCE, a cascading collapse struck many Bronze Age civilizations. The Hittites were not spared. Pressure from Sea Peoples, internal rebellion, famine, and geopolitical stress brought their empire to ruin. Cities burned. Palaces crumbled. Workshops fell silent.

With the collapse came a scattering of knowledge and opportunity. Blacksmiths fled or were captured. Iron techniques, once sealed within empire walls, slipped into rival lands and among wandering groups. The secret that once belonged to a single power began to move. Across the Near East and into Europe, societies brushed away the embers of disaster and discovered iron within the ashes. The monopoly ended. The Iron Age began.

Kings, Diplomats, and the Politics of Metal

Iron altered relationships between states. Royal letters preserved on clay tablets, many recovered from the Hittite capital of Hattusa, offer glimpses into diplomatic exchanges focused on metal. Kings requested iron gifts to demonstrate loyalty, sometimes pleading politely, sometimes demanding as equals. A ruler's prestige rose when he possessed iron. It became a bargaining chip for peace, marriage alliances, and regional influence.

Other kingdoms pursued iron desperately. Mycenaeans sought warriors equipped for new threats. Assyrians, destined to become one of the ancient world's fiercest powers, embraced iron as a foundation of expansion. Controlling iron became a cornerstone of political power.

Technology as a Force of Change

Knowing how to work iron meant mastering fire itself. Once extracted from ore, iron required refinement: heating, folding, and hammering to remove slag and increase strength. Heat treatment—cooling the blade quickly or slowly—changed hardness, sometimes making a weapon too brittle, sometimes strengthening it perfectly.

This knowledge, learned through practice and memory rather than chemical theory, allowed the Hittites to craft reliable military iron. Techniques passed from master smith to apprentice, shaped by experience. Written records lacked technical details. Almost everything depended on lineage and craft tradition.

When that tradition dispersed after the empire's fall, iron knowledge spread rapidly. Blacksmithing no longer belonged to a single capital. It belonged to every village furnace willing to experiment. Innovation moved from state control into local craft—and that democratization reshaped the ancient world.

The Shift That Resculpted Armies

As iron production grew elsewhere, it no longer remained a luxury. Iron tipped the spears of common soldiers. Iron axes placed power in the hands of farmers pressed into service. Iron armor protected troops previously left vulnerable. Societies that had relied on elite bronze chariot warriors—expensive to outfit and train—now fielded larger numbers of iron-equipped infantry. The nature of warfare shifted. The Hittites had lit the match. Others built the firestorm.

In time, Assyria mastered iron mass production and conquered lands the Hittites once ruled. Greece and later Rome refined forging techniques. New tactical formations appeared. Battles depended not just on wealthy aristocrats but on citizen soldiers, igniting social change. A blade is never just metal. It is the beginning of a different kind of world.

Religion and Ritual in the Age of Iron

Even as iron moved from sacred ritual to daily life, its divine associations lingered. Hittite texts link iron to the heavens, perhaps reflecting the earliest iron objects formed from meteorites. Storm gods and war gods received iron relics as offerings, suggesting a belief that the metal bridged earth and sky.

When common soldiers carried iron into battle, they may have believed they carried divine strength. Every clang of a blade could echo thunder. Legends outlast their makers. When the Hittite Empire faded, the spiritual power of iron did not fade with it.

What the Hittites Began

Archaeologists continue to unearth Hittite forges and tools, charred remains of workshops where metallurgists once guarded secrets that defined an age. The evidence is incomplete but undeniable. The Hittites accelerated humanity's shift from bronze to iron.

They pushed beyond limitations. They experimented where others hesitated. They transformed a difficult metal into a weapon capable of overwhelming enemies and reshaping society.

Their empire did not survive the crises of the late Bronze Age. Yet the metal that helped fuel their rise survived everywhere.

The Weapon That Outlived the Empire

Standing on that ancient battlefield once more, one can imagine shattered bronze strewn across the dust while iron blades remain intact. A symbol of what comes next. A preview of changes the world could not yet understand.

Iron promised independence from distant tin mines. It shifted wealth from a few coastal traders to countless inland blacksmiths. It changed armies, economies, and belief systems. It empowered those who wielded it and humbled those who ignored it.

The Hittites sparked a transformation so vast that their own collapse could not halt it. The Bronze Age ended not with diplomacy or disaster alone, but with the rise of a harder truth. When iron entered war, the world itself changed allegiance.

Chapter 9
The Sun Forged in Metal – Inca Alloys and the Empire of Gold

High above the clouds, where the thin air tastes sharp and the horizon glows with first light, priests carry a golden object toward a waiting crowd. The morning sun rises over the spine of the Andes, its rays striking the polished surface until it blazes like a second dawn. The object seems alive, breathing light across stone terraces and silent peaks. To the people gathered here, this is not merely treasure. This is sunlight captured, shaped by human hands, and returned to the sky in worship. Metal is a bridge between worlds. In the Inca Empire, gold is the sweat of the sun, silver the tears of the moon, and copper the lifeblood of the earth. In this belief, metallurgy becomes devotion, economy, and power intertwined.

Across the Andes, centuries before the arrival of conquistadors, skilled metalworkers forged a civilization's identity in brilliant metals and complex alloys. Their mastery, quietly developed in highland workshops and coastal valleys, was as sophisticated as anything seen in the Old World. Yet its purpose was profoundly different. In this empire, metal measured spirit more than wealth.

Sacred Metals in a Vertical World

The Andes offered a landscape of extremes. Snow-capped mountains fed rivers that carved through valleys and deserts, creating narrow strips of livable terrain. The people who flourished here learned to draw life from steep fields and hidden mines. Gold and silver came from high elevations, extracted from veins that shimmered in jagged rock. Copper came from vast deposits in lower zones, linked to the earth's creative power.

The Incas absorbed and expanded traditions from cultures that preceded them. The Moche of Peru's northern coast were famed for expressive portrait vessels and ritual metal masks. The Chavín culture of the highlands blended animals, humans, and gods in forms hammered into gilded objects.

These societies set the foundations for a metallurgy that emphasized ritual over warfare. Bronze swords never dominated the Andes as iron blades did in the Old World. Instead, metals became material prayers. To possess a gold ornament was to carry a fragment of the divine. To lay a silver idol in a mountain shrine was to speak directly to the moon. Andean metallurgy was, first and foremost, spiritual.

Tumbaga: A Hidden Masterpiece

One of the most ingenious achievements of Andean metallurgy was the alloy known today as tumbaga, a blend of gold and copper sometimes enhanced with other metals. When first removed from a mold, tumbaga could appear duller than pure gold. But artisans used a technique called depletion gilding, bathing the alloy in acidic solutions from natural plant extracts or mineral-rich soils. This process leached copper from the surface while preserving gold, leaving a thin, radiant skin.

The result confused early Europeans who measured wealth by purity. Tumbaga dazzled like solid gold yet contained far less precious metal inside. But it was not deceitful. It reflected a belief that brilliance mattered more than weight. A divine symbol needed only to appear like the sun. Its radiance carried power.

For Andean metallurgists, the goal was not to hide value but to create beauty and meaning from multiple elements. Just as the empire blended cultures, tumbaga blended metals to create something new.

Workshops of Empire

Across the Inca realm, skilled metallurgists—known in Quechua as *urumi*—worked in imperial centers such as Cuzco and at state-run sites in conquered territories. Their expertise was protected and prized. Some were captured from communities known for their skill and required to serve the empire as specialists, much like artisans relocated under the Hittites or craftsmen of Egypt's temples.

These workshops produced ornaments, ritual objects, agricultural tools, and decorative fittings for temples and palaces. They hammered thin sheets of gold into walls that shimmered like morning light. They shaped silver into life-sized figures of llamas and maize plants, symbols of prosperity offered to the gods. Even ordinary copper tools, cast in standardized forms, enabled terrace farming and construction that sustained millions.

In the Andes, technical skill was inseparable from sacred duty. Whether a metalworker hammered a nose ring for a warrior or a ceremonial blade for a priest, the object carried cosmological significance.

Gold for the Sun, Silver for the Moon

Religion infused every metal. Inti, the sun god, received gold shaped into discs and panels, his temples glowing like sunrise. Mama Quilla, the moon goddess, received silver offerings that sparkled in night rituals. Copper marked the power of Pachamama, the earth mother. By transforming metal into devotional gifts, the Incas believed they balanced cosmic energies that sustained agriculture and life.

This symbolism extended into social roles. The Inca emperor, regarded as the son of the sun, covered himself in gold regalia that projected legitimacy. Nobles adorned themselves in metal jewelry that signaled their closeness to divine authority. Even death required metal. Elite mummies wore gold masks so their spirits would be recognized by gods in the afterlife.

Some metals were so sacred they never circulated as currency. The empire used labor and food as the foundation of exchange. Metal measured devotion, not market value.

Tools of Power, Not War

Unlike the iron-driven militaries of Eurasia, the Inca armies relied on scale, organization, and engineering rather than metallurgical weapon superiority. Western chroniclers, expecting steel swords and bronze armor, expressed confusion at the relative scarcity of metal in warfare.

The empire's greatest weapons were roads, bridges, and logistics, supported by metal tools. Copper-arsenic bronze formed axes and chisels used to carve terraces into mountainsides and build storage houses for grain. Sling stones and clubs still dominated battles, but behind every soldier stood an infrastructure forged by metal.

Metals enabled the empire to move, feed, and unite its people. The brilliance of a gold temple was possible because copper tools fed the workers who built it.

Knowledge Passed in Silence

Unlike civilizations that documented industry in writing, the Inca recorded information using knotted cords called quipus. These served administrative needs but left metallurgical practices largely unrecorded in decipherable form. Modern scholars therefore reconstruct ancient techniques through artifacts, excavated furnaces, and experimental replication.

The absence of written manuals makes their success even more remarkable. High-altitude furnaces demanded advanced understanding of heat and airflow. Craftspeople compensated with bellows and carefully designed chimneys. Quartz and clay molds allowed for detailed casting using lost-wax methods borrowed from earlier Andean cultures. Surface treatments created textures and colors unmatched elsewhere at the time.

The sophistication visible in the artifacts speaks for the unrecorded brilliance of those who made them.

The Metal Economy of an Empire

The Incas did not sell gold. They assigned it to temples. They did not mint silver as coins. They offered it to the moon. Their metal economy reinforced hierarchy and unity. Commoners produced food and textiles. The state redistributed resources through an immense labor tribute system. Metal reinforced prestige and reminded every region that Cuzco sat at the center of the world.

The state also used metallurgical gifts to win loyalty. Conquered leaders who submitted to Inca rule received ornaments and tools, visible proof of inclusion in something larger. Metal turned subject communities from outsiders into partners, even if reluctantly. Refusal to acknowledge imperial gifts often led to conquest. In this way, alloys became diplomacy. Gold was soft in hand but firm in politics.

The Conquistadors and the Clash of Values

When Francisco Pizarro arrived in the 1530s, he encountered palaces sparkling with metalwork beyond European imagination. The Spaniards saw treasure. The Incas saw gods. Misunderstanding led to destruction. Temples were stripped. Sacred objects were melted into ingots. Master artisans were enslaved or killed. Techniques perfected over millennia vanished abruptly.

The loss was not only material but cultural and spiritual. Objects that once connected sky to earth became currency. Quipu-keepers died with their knowledge. Gold stopped shining for the sun and instead filled ships bound across the sea.

Where Spanish records describe plunder, archaeology describes silence—a sudden gap where innovation once thrived.

What Remains Beneath the Mountain Light

Despite devastation, Andean metallurgy refuses to disappear. Buried beneath centuries of soil, metalwork emerges from tombs and temple ruins, telling stories the conquerors never understood. Museums display shimmering masks, miniature llamas, ceremonial cups, and ornaments shaped like flowers offered to gods of rain and growth.

Modern metallurgists test these artifacts and marvel at alloy recipes balanced with precision and daring. They replicate depletion gilding and rediscover ancient brilliance. Archaeologists map workshops and reveal how labor and artistry structured urban life. Descendants of the Inca reinterpret metals in festivals that still honor sun and moon.

The story did not end with conquest. It continues in identity, memory, and shine.

A Civilization Reflected in Its Metal

Stand again on that mountaintop. The sun has risen fully now. Gold still catches its fire, just as it did centuries ago. Empires fall. Temples crumble. But metallurgy endures. In every gleaming artifact, one sees high mountains, fertile valleys, and the people who shaped them with belief as much as skill.

The Inca and their predecessors forged metals not to dominate land with swords but to connect heaven and earth. Their alloys reflected gods, legitimized rulers, beautified life, and preserved harmony. They built an empire on terraces and prayer, shining with the colors of the sun and moon.

Metals did not merely support their world. They expressed what that world believed: that the divine could be shaped by human hands, and that in the glow of gold, one might glimpse eternity.

Chapter 10
The Architect's Metal – Bronze and the Foundations of Greek Engineering

Blinding sunlight washes over a rising temple as workers hoist an enormous bronze-sheathed door into place. Its surface reflects daybreak in dazzling flashes, turning the structure into a beacon of ambition and reverence. Thick bronze hinges, shaped and polished with quiet artistry, fit so perfectly against their stone sockets that the giant door glides into position with surprising ease. From the flick of a lever to the swing of the door, the craftsmanship whispers a truth often overlooked. In ancient Greece, bronze was not merely the metal of warriors and statues. It was the hidden skeleton of innovation.

For centuries, Greek engineers and artisans used bronze to turn impossible ideas into practical achievements. In temples, ships, theaters, workshops, and inventors' studios, bronze shaped the infrastructure of a civilization that prized intelligence as a path to power.

Beyond Spears and Helmets

Greek bronze is often imagined in the hands of heroes: the spear of a hoplite, the helmet of a Corinthian warrior, the blade of Achilles gleaming in epic poetry. The metal indeed defined martial identity. Bronze weapons sharpened into lethal edges. Bronze armor gave hoplites the resilience to form tight phalanxes that dominated early classical warfare.

Yet the Greeks saw more in bronze than battle. They valued its mechanical strength, corrosion resistance, and ability to hold detailed shapes through casting. Where some cultures confined bronze to ritual goods, the Greeks made it a tool of thought.

Artisans crafted bronze mechanisms long before the Industrial Age imagined gears and pulleys. Engineers and philosophers touched bronze to bring theories into the world of action. Without bronze, Greek innovation would have remained imagination alone.

Bronze as Architecture's Unsung Hero

Some of the most impressive Greek temples survive because bronze made their architecture possible. The massive stone blocks of the Parthenon and other structures required precision beyond what muscle alone could achieve. Bronze clamps and dowels locked stone pieces into stability. Hinges, pins, and fasteners—many later melted down—served as the connective tissue behind the grandeur.

Bronze doors signaled sanctity and civic pride. Important temples displayed entrances covered in sheets of gleaming alloy, strong enough to resist weather and elegant enough to designate divine presence. The bronze elements of a temple acted with quiet authority, helping the structure endure earthquakes, storms, and centuries of devotion.

Even water management relied on bronze. Aqueducts and fountains used bronze valves and spouts to regulate flow. Sacred sites such as Delphi used bronze fittings hidden beneath stone walkways to carry fresh water through channels pilgrims never saw.

Ships That Carried Empires

Greek naval power rose not from wood alone. Bronze played a decisive role in the design and performance of triremes and other warships that secured both trade and territory.

Bronze rams—deadly protrusions at ship bows—gave Greek navies tactical advantage. In battles like Salamis, Greek rowers drove their vessels forward with precision, slamming bronze-tipped hulls into enemy craft. One well-placed strike could splinter wooden planking and send rival warriors into the sea.

Bronze nails, pins, and reinforcements protected ships from splitting under stress. Even rigging benefited from bronze pulleys that let crews maneuver sails quickly. Greek seafaring dominance was as much a triumph of metal engineering as of strategy.

Merchant vessels also carried bronze goods, from raw copper and tin to finished tools. Wherever Greek trade spread, bronze followed.

The Greek Mechanists

Greek thinkers approached the world with an engineer's curiosity. They experimented, measured, and built. Bronze gave shape to those investigations.

The Antikythera mechanism, found in a shipwreck near a small Greek island, is among the most extraordinary examples. Dating to the second century BCE, it contains dozens of bronze gears arranged with astonishing precision. It functioned as a mechanical computer predicting lunar phases, eclipses, and celestial cycles.

Its existence proves that Greek metalworkers could cut gear teeth so uniform that the device still amazes modern science. Bronze, cast and filed with patience, became a vehicle for intellectual exploration.

Inventors such as Hero of Alexandria later developed bronze-driven automata, water organs, temple doors that opened through hidden mechanisms, and early steam devices. Though Hero lived in the Hellenistic period under Roman rule, his lineage grew from Greek thinkers who turned ideas into machines.

Even Greek theater benefited. Stage mechanisms, moving platforms, and sound devices used bronze components designed to endure repeated use. Drama itself was engineered from metal as much as from poetry.

Bronze in Daily Work and Craft

Greek households and workshops depended on bronze for countless practical tools. Farmers used bronze sickles to harvest grain. Carpenters employed bronze chisels and drills to shape wood. Metalworkers used bronze anvils and hammers to strike other metals.

Artisans relied on bronze molds to mass-produce items ranging from jewelry clasps to standardized market weights. Through these tools, bronze acted as a force multiplier. Every strike shaped not only an object, but the economy that depended on it.

Bronze coins became currency across many Greek regions. Silver dominated high-value transactions, but bronze coinage allowed everyday citizens to participate in trade. Bronze helped democratize the market.

Trade, Resources, and the Price of Innovation

Bronze required a delicate balance of geography and politics. Copper deposits existed in Cyprus and mainland Greece. Tin, however, was scarce and often traveled from distant lands, possibly northern Europe.

Securing tin meant diplomacy, alliances, and sometimes conflict. Athens and Corinth invested heavily in naval power to ensure raw materials reached their ports. Control of the seas meant control of bronze.

Greek colonies helped secure ore sources, weaving metallurgy into cultural expansion. Every city-state's relationship with bronze reflected its ambitions.

This dependence also created vulnerability. Interrupt tin supply and bronze production faltered. Greek engineering could stall from a single broken trade route.

When Iron Stepped In

Iron's rise did not erase bronze immediately. Even during the Iron Age, bronze retained roles in fine craft, armor fittings, art, and corrosion-resistant engineering components.

Iron carried one new advantage: abundance. Iron ore was easier to obtain. As ironworking techniques matured, production surged. Iron tools could be made for more people, more cheaply.

By the classical period, Greek armies bore iron weapons. Iron ploughs cut deeper into rocky soils. Iron nails held buildings together more firmly.

Bronze remained noble and precise. Iron became common and transformative. Greek power shifted with the metal that armed it.

Legacy Cast into History

Walk among Greek ruins and bronze no longer gleams from every hinge and fitting—most were scavenged long ago. Yet the stones still show where metal once reinforced greatness.

Visit the remains of ship sheds in Piraeus and imagine bronze-edged warships sliding into the sea.

Look at the Antikythera mechanism and witness genius hidden in corroded gears.

Bronze is the quiet storyteller of Greek advancement. It allowed thinkers to measure the heavens, architects to anchor stone, sailors to rule the sea, and engineers to awaken movement from stillness.

The Greeks asked what the human mind could accomplish and answered through alloys and invention. Bronze carried those answers into reality.

Even in an age dominated by iron, bronze retained honor—a reminder that civilizations rise not only through conquest, but through craft. Greek bronze built foundations strong enough for ideas to climb. In every hinge that still turns and every artifact that reflects sunlight after millennia underground, bronze continues to reveal Greece's truest marvel: a belief that knowledge, given form, becomes power.

Chapter 11
Armor of a Thousand Scales — Japanese Steel and the Art of the Samurai Sword

Night hushes the mountain village, but in one wooden smithy the silence breaks again and again with the sharp chime of hammer striking steel. Each blow scatters sparks that briefly illuminate the swordsmith's focused expression. He bends over a glowing billet of metal, folding it and hammering impurities into submission. Tamahagane steel, born from black iron sand and fire, transforms with every strike. The workshop fills with the smell of charcoal and the pulse of flame. When the blade reaches its moment of truth, he plunges it into water with a hiss that rises like a dragon's breath. Steam veils the room. Lantern light catches the emerging hamon line, its ghostly shimmer running along the edge as though the steel itself remembers the fire. Here, technology and spirituality share the same heartbeat, and the steel becomes more than a weapon. It becomes a soul.

The Steel of the Island

Japan's path to steel mastery began with geography. Unlike regions blessed with high-grade iron ore, Japan possessed iron sand, tiny grains washed from volcanic rock and carried into riverbeds and coastal pockets by mountain streams. Harder to smelt than pure ore, the sand demanded furnaces capable of intense heat and patient refining.

Japanese metallurgists responded with the tatara furnace, a long clay structure fueled by charcoal and fed continuously with alternating layers of iron sand and fuel. For three days and nights, workers stoked the fire tirelessly. Inside, chemical reactions transformed the sand into blooms of steel with varied carbon content. From this rough mass, swordsmiths selected only the finest tamahagane, judging quality by sound, sparkle, and feel rather than written formula.

This scarcity of pure materials shaped everything that followed. Where other civilizations produced iron in large, ingot-based systems, Japan refined smaller amounts with immense labor. The sword that emerged was not a common tool. It was the apex of craft in a metal-poor land.

Folding Fire into Purity

Once chosen, tamahagane endured ritual as much as refinement. The smith heated it, hammered it flat, folded it over, and repeated the process many times. Each fold drove out slag, aligned the grain, and distributed carbon more evenly throughout the blade. A skilled smith balanced hardness and flexibility through instinct and experience.

To the untrained eye, folding might seem excessive or decorative. In practice, it was a necessity of the raw material. Because the initial steel contained impurities and inconsistent carbon, folding healed the metal. It turned weakness into layered resilience.

The result was a weapon composed of thousands of micro-thin layers, each contributing to strength. Under a microscope, a finished blade reveals patterns like flowing scales, a metal with memory and movement.

The Secret of Differential Hardening

The most distinctive feature of the samurai sword is the hamon, the wavy boundary between the flexible spine and the razor-hard edge. This line is no ornament. It is the visible signature of differential hardening, a process in which smiths coated the back of the blade with insulating clay, leaving the edge exposed.

When plunged into water or oil during quenching, the exposed edge cooled rapidly, locking carbon into a hard martensitic structure. The clay-covered spine cooled more slowly, remaining tough and bendable. A blade emerged that could cut through armor yet flex under pressure, part bone, part muscle. Its form echoed the duality of nature embraced in Zen philosophy.

The hamon offered more than function. It offered identity. Each smith crafted patterns unique to his school, straight lines, waves, peaks, or flickering lightning shapes. These designs became signatures in steel.

Legends in Steel: Masamune and Muramasa

Swordsmiths in Japan earned reputations equal to those of poets and generals. Two names rise above the rest, Masamune and Muramasa.

Masamune, active during the late Kamakura period, forged blades known for beauty and refined performance. His swords cut cleanly without deforming, embodying restraint and spiritual harmony. Legends claim they brought peace to those who wielded them.

Muramasa, several generations later, forged blades said to be fierce and unyielding, sharp enough to cut with the faintest motion. Stories describe his swords as cursed, bloodthirsty weapons that demanded violence. In reality, these myths reflect cultural anxieties more than metallurgy. But they show how deeply the Japanese people saw personality in metal, as though swords carried the temperament of their maker.

The rivalry is poetic rather than literal, yet it demonstrates that Japanese craftsmanship was judged not only by performance but by moral and spiritual philosophy.

The Sword as Soul

By the rise of the samurai class, the sword had ascended from weapon to symbol. Confucian principles and Shinto beliefs merged to assign spiritual significance to blades. A sword was a container for honor, courage, and ancestry. When a samurai died, his sword often accompanied him into burial as proof of his identity.

Ceremonies surrounded every stage of the sword's life. Blades were blessed at shrines. Swordsmiths prayed before forging. Warriors treated swords like family members and passed them down across generations. To lose a sword was to lose face.

Even during the long peace of the Tokugawa shogunate, swords remained potent symbols of authority. Samurai wore daishō, paired long and short swords, as visible proof of their rank. The blade might rest quietly, but its meaning endured.

A Different Metallurgical Road

Japan's distinct sword culture emerged not simply from artistry but from necessity and isolation. Unlike China, which advanced iron production to monumental scale, Japan focused on small-batch refinement. Unlike Europe, which evolved armor and swords through constant competition, Japan spent long periods closed to foreign influence, locking its metallurgy inside a cultural vault.

Trade restrictions slowed the adoption of foreign steel and new techniques. Japan relied on homegrown mastery instead. This isolation preserved tradition but limited industrial progress. When firearms arrived in the 16th century, the sword did not disappear, but its battlefield supremacy faded.

Still, tradition defended the forge. Even as muskets reshaped warfare, swordsmiths continued refining techniques that had taken centuries to perfect.

The Edge of Myth and the Truth Beneath

Stories claim a katana can slice through stone pillars or cut a rifle barrel in half. These feats are exaggerations born from admiration. Steel has limits. Even the most perfectly tempered blade can chip or break.

Yet the reality remains impressive. The hardness of the edge and the flexibility of the spine give the katana remarkable balance. It can score armor, sever bone, and endure shocks that would break simpler blades. Modern metallurgical testing confirms the precision smiths achieved over carbon structure, science walking the same road that craft had paved long before. Fact and legend blur because truth already feels legendary.

Secrecy as Both Strength and Weakness

Swordsmithing knowledge passed through families, guarded as fiercely as clan honor. Masters taught apprentices only when trust was complete. Techniques varied by province, and rivalries discouraged sharing.

This secrecy preserved quality. But it also limited adaptation when new technologies challenged the sword's dominance. The samurai ethos valued continuity over reinvention, and metallurgy reflected that choice.

As Japan modernized in the late 19th century, the sword's role shifted again from battlefield tool to cultural symbol. The Meiji government banned samurai privileges, yet the sword found new life in ceremony, martial arts, and heritage. Traditional forging survived because it represented identity as much as craft.

The Legacy of a Blade That Remembers

Today, Japanese swordsmiths still fire tatara furnaces, still fold tamahagane, still bow before the forge. Their work is protected as an intangible cultural treasure. Museums display blades with reverence. Martial artists train with replicas that honor form and philosophy.

The sword remains a bridge between eras, a reminder that human ingenuity once combined religion, science, and discipline into a single object sharp enough to command history.

Look again into that mountain smithy. The steel has cooled. The hamon glows like a silver horizon. A sword lies ready, balanced not only in structure but in meaning. It is both weapon and wisdom, the embodiment of a culture that shaped metal into character.

Japan's steel endures because it is more than carbon and iron. It is identity shaped by fire.

Chapter 12
The Inferno Beneath the Desert — Persian Furnaces and the Fire of Innovation

Heat radiates off the harsh desert plain where cliffs rise like ancient guardians. Workers toil under a blistering Persian sun, their skin shining with sweat and grit. In front of them, a towering mud-brick furnace roars alive. Wind funnels through carefully placed vents, feeding the charcoal with invisible breath. Flames leap upward as if reaching for the sky, licking the ore suspended inside. Metal glows in a blistering orange bloom, a miniature sun held captive beneath barren stone. The craftsmen move quickly, guided by instincts passed down through generations. They know exactly when to add fuel, when to adjust the airflow, and when to tap the molten heart of the furnace. Every decision, every moment, carries the weight of empire.

Here in the ancient lands of Iran, Persian metallurgists reshaped the material foundations of power. They refined iron and steel with skill that few could match, constructing tools and weapons that would build dynasties, command trade routes, and challenge the greatest armies of the ancient world.

The Crucible of Crossroads

Geography made Persia a place where ideas converged. Mesopotamian societies lay to the west. The Indus Valley and Central Asia brought influences from the east. Nomads traversed steppe routes to the north, and Arabian trade surged from the south. At this crossroads, Persia absorbed knowledge and transformed it, forging a metallurgical identity that blended creativity with practicality.

Long before iron crowned the empire, the people of Luristan in the Zagros Mountains mastered bronze craftsmanship. The artifacts they left behind—ornate daggers, horse fittings, and ceremonial standards featuring spirals and divine figures—demonstrate artistic sophistication and technical courage. Their understanding of alloy composition laid a foundation for future advances in iron and steel.

By the rise of the Achaemenid Empire in the sixth century BCE, Persia was ready for a new age of furnace-driven innovation. Cyrus the Great and his successors needed metal that supported infrastructure and warfare. Iron became the steel spine supporting an empire stretched from the Aegean Sea to the Indus River.

Climbing the Furnace Toward Steel

The key to Persian success lay in furnace design. Taller mud-brick structures captured heat while directing airflow through carefully positioned side vents. Persians recognized that air was not merely a helper but a reagent, one that could radically alter the quality of metal.

Bellows helped when nature did not cooperate, but in windy desert regions, smiths cleverly exploited their environment. Natural gusts fed the flames so efficiently that the furnace seemed alive. Today, archaeologists find slag heaps near abandoned workshops, silent records of immense industrial activity.

Iron extracted through bloomery smelting emerged mixed with slag and required repeated reheating and hammering to compact and refine it. Persian smiths improved these steps by manipulating carbon content through charcoal contact. Though they lacked chemical theory, they possessed generations of experimental wisdom. They coaxed iron into becoming steel.

Steel for Conquest

The Persian military turned these advances into geopolitical strategy. Elite infantry, including the famed Immortals, wielded iron or steel-tipped spears and arrows that outlasted bronze-equipped opponents. Armor incorporated rows of metal scales stitched into fabric, balancing defense with mobility.

Cavalry and chariot forces carried weapons forged in the fires of the Zagros.

Tribute flowed through the empire in the form of raw metals, finished blades, and decorative fittings that displayed imperial wealth. Metallurgy supported both diplomacy and domination.

This strength mattered deeply during the Greco-Persian Wars. At Marathon, Thermopylae, and Salamis, the soldiers of Darius I and Xerxes I brought the best metallurgy the East could offer. Greek writers—enemy observers though they were—recorded Persia's superior materials with grudging respect.

To support this war machine, Persia needed not only weapons but infrastructure. Metal tools cut stone for the Royal Road, which stretched more than 1,600 miles and enabled communication at astonishing speed. Irrigation canals, monumental palaces like Persepolis, and fortified cities all relied on iron tools that multiplied the efficiency of labor.

Steel forged power. Tools sustained it.

The Enduring Fire of Bronze

Even as iron rose, bronze retained a vital presence in Persian culture. The bronzes of Luristan show that artistic metallurgy remained central to Persian identity. Curved animal forms, divine figures atop standards, and elegant chariot fittings reveal that metal in Persia was never just functional. It was spiritual. It was political. It was the voice of culture hammered into shape.

Trade Paths of Fire

Persia did not operate in isolation. Iron from the Caucasus, copper from Oman and Arabia, tin from Central Asia, and specialized steels from India moved along networks the empire guarded fiercely. Persian workshops exported tools, blades, and artwork that traveled the early Silk Road and coastal trade routes.

Trade spread Persian innovation as effectively as conquest. Ideas moved with caravans. Craftsmanship migrated with captured or hired smiths. Workshop secrets traveled farther than borders.

Innovation as Strategy

Persian rulers understood that technology was power. Better metal tools improved farming yields, which fed the army. Better weapons discouraged rebellion. Better infrastructure held provinces in unity.

Even enemies learned from the Persians. Greek manufacturers adapted techniques encountered in war. Later empires implemented Persian metallurgical strategies as readily as they copied Persian governance. Fire became the empire's silent strategist.

Collapse and Continuity

The fall of the Achaemenid Empire to Alexander the Great in the fourth century BCE shifted political authority but did not extinguish the smelting fires. Metallurgical knowledge endured, carried by craftsmen into Hellenistic and later Islamic workshops. Their techniques contributed to breakthroughs in crucible steel and patterned metal centuries later.

Civilizations may collapse, but innovation survives through the hands that remember.

The Legacy Beneath the Sand

Across Iran today, archaeologists uncover furnace fragments and slag that testify to a relentless pursuit of metallurgical excellence. Researchers examining these remains find evidence of efficient heat management, carbon-control techniques, and sophisticated large-scale production.

Every discovery reshapes our understanding and confirms Persia's vital role in shaping the ancient world's technological future.

Stand again beneath the desert sun. The furnace is roaring. The air trembles with heat. Molten metal gleams like a fragment of the sun waiting to take form. The workers are not only hammering weapons. They are forging history.

Persian metallurgy turned fire into policy, ore into empire, and innovation into power. Its flames did not merely temper steel. They tempered civilization.

Chapter 13
The Alchemists' Metal – Zinc, The Invisible Element that Changed Everything

Early metallurgists stare at a stubborn crucible as blue-white fumes curl upward and vanish. The fire is hot, the charcoal rich, the copper ready. Yet the pale vapor refuses to settle into a shining bead. It escapes through hairline gaps and disappears into the night air, leaving a ghostly glaze on the clay. They know the vapor is metal in an elusive form, because it changes the copper when enough of it finds its way into the melt. The result is a warm, golden alloy that cuts cleaner and resists the sea. They do not have a word for the fugitive element, only a growing sense that an unseen partner hides inside their furnaces, changing the world without showing its face.

The Missing Ingredient That Was There All Along

Long before anyone wrote "zinc" on a tablet or page, metalworkers stumbled into brass by accident. In West Asia and South Asia, copper ores often sat near zinc-bearing minerals such as smithsonite or calamine. When ancient smelters roasted these mixed ores with charcoal, a share of zinc vapor seeped into the copper melt and dissolved. The product looked like gold to the untrained eye, though it behaved like tougher copper. It became prized for pins, buckles, mirrors, bowls, and fittings that resisted corrosion and rang with a clearer tone.

Artisans felt the difference in hammer and file. Brass took a crisp edge on decorative reliefs. It flowed into molds with fewer defects than pure copper. Temple bells held steadier voices. Still, the makers lacked a reliable recipe. The invisible metal arrived when it wished and vanished when the furnace crew pushed too hot or too fast. Everyone sensed a rule at work. No one could write it down.

Why Zinc Was So Hard to See

Copper offers the smith a generous window. Heat it, and it softens before it melts. Tin behaves in a similarly forgiving way, which is why bronze spread so early. Zinc lives on a narrower ledge. Its boiling point sits well below the temperature of a vigorous charcoal fire. The moment a furnace climbs into a zone that smelts copper, zinc turns to vapor. If the vapor meets oxygen, it burns into a white powder and drifts away. If the vessel leaks, the element is gone before it can join the copper. The problem is not scarcity. The problem is control.

For centuries, the only practical path was indirect. Let zinc vapor form inside a sealed space packed with copper. Let it soak into the metal before it finds air. The invisible worker did not need a name to earn a place on palace ledgers.

Early South Asian Know-How

Evidence from the Indian subcontinent suggests a turning point. By the early first millennium CE, and likely earlier in some districts, furnace workers learned to manage zinc vapor on purpose rather than by accident. At mining centers such as Zawar in Rajasthan, they roasted calamine and ran distillation stacks of ceramic retorts that sloped downward into receivers. Inside each vessel, zinc ore and charcoal shared a sealed chamber. As the ore reduced, zinc vapor formed and traveled through the neck of the retort into a cooler zone, where it condensed as liquid metal or as spongy crystals that could be scraped and remelted.

Workers judged heat by flame color, air draft, and the sound of the draw inside the flues. They did not need modern gauges to make choices with repeatable outcomes.

Once you hold zinc as a metal, brass becomes a plan rather than a prayer. Foundries can weigh portions, charge crucibles with known ratios, and produce alloys for specific jobs. High-zinc brass carries the warm yellow merchants want to display. Lower-zinc mixes strengthen hardware for shipyards and gates. The same distillation logic feeds medicine and pigment, since zinc compounds brighten glazes and soothe the skin. Industry branches outward from a single disciplined fire.

China's Experiments and Parallel Paths

Far to the northeast, Chinese metallurgists pursued their own routes to the same prize. Early brasses formed from mixed ores in the Warring States and Han periods, useful for mirrors and fittings but prone to variation. Over time, workshops refined sealed-crucible methods and learned to coax more zinc into copper without losing it to the flame. Later, technical treatises recorded controlled heating, careful sealing, and timed decanting to keep zinc from burning off.

China's long tradition of cast copper-alloy mirrors shows what stable brass chemistry meant. Flatness and reflectivity demand patience in polishing, but also an alloy that behaves consistently. When zinc is under firm command, the artisan can focus on the surface rather than fight the metal beneath.

The Mediterranean's Half-Light

In the Mediterranean, workshops had made brass by cementation for centuries. They packed copper with crushed calamine inside sealed crucibles and heated the charge for long hours. Zinc vapor formed inside the pot and diffused into the copper. The method could yield fine brass, but it never revealed the ghost directly. Most makers did not see zinc as a separate metal, only as a behavior that copper could exhibit when cooked with the right stones.

As a result, control stayed imperfect, and the alloy's reputation rose and fell with each batch. Some cities became known for reliable brass wares, while others sold bright objects that dulled too soon. Guild secrets hardened, and recipes became guarded songs.

Why Zinc Mattered to Armies and Kings

A coin that lasts keeps a promise. A hinge that resists the sea keeps a fleet ready. A trumpet that holds its tone keeps an army in step. Zinc quietly strengthened these bonds of state.

High-zinc brass endured salt spray better than copper alone, which reduced maintenance in naval stores. Armor mounts stayed sound in humid climates. Precision parts for crossbows, locks, and surveying tools wore more evenly. Brass stood in for gold when budgets were tight, yet authority needed to shine.

Control of zinc production created new points of leverage. A court with steady distillation could pay workers with reliable low-denomination coinage. A port that sold consistent brass sheet could draw shipwrights from abroad. States wrote treaties about silver, iron, and timber. Zinc rarely appeared in inscriptions, yet it helped decide who brought their fleets home intact.

Secrecy, Scarcity, and the Price of Knowledge

Every breakthrough invites a closed door. Furnace builders who mastered retorts and receivers guarded their methods. A poorly sealed joint could waste weeks of labor. An air leak at the wrong moment could scald a crew with zinc oxide smoke and ruin a charge. Families taught their best tricks only to kin. Regions with the right clays for retort necks kept their sources quiet.

Secrecy slowed the spread of the technology, yet it also protected quality long enough for some centers to build wealth. When knowledge leaked, it often traveled with artisans, not scrolls. A potter who could throw heat-resistant ceramics, a smelter who read draft like a sailor reads wind, a foreman who listened for the moment a retort begins to sing—these were the vectors of change.

The Long European Delay

Europe's relationship with zinc followed a winding road. Roman and medieval craftsmen made good brass by cementation. They recognized that certain "calamine" stones added a golden tint to copper. They refined workflows around sealed pots and long soaks. But isolating metallic zinc itself remained elusive. When heat climbed to the level needed to reduce the ore, zinc kept boiling away.

By the late medieval and early modern periods, travelers carried stories of eastern methods back to European workshops. Retort designs gradually appeared that could capture zinc vapor before air destroyed it. Once zinc could be held as a metal, brass mills flourished in new ways. Uniform sheet and wire supported clockmaking, scientific instruments, buttons, and cartridge cases. The invisible ally finally had a name, and with the name came standards.

Isotopes That Follow the Money

Modern science has given historians a new lantern for dark corners. Zinc, like other elements, occurs in several stable isotopic forms. Ores from different geological settings carry distinct isotopic fingerprints. When researchers analyze an ancient brass artifact, the zinc's isotopic pattern can hint at its source. Combine that with trace elements and known mining districts, and trade routes begin to emerge on the map.

These signatures have traced paths from Central Asian or Iranian ore belts into South Asian refineries and onward to coastal markets. They have linked European objects to specific mining regions and shown how shipments shifted when wars closed passes or when new ports rose under ambitious rulers.

Industries Quietly Rewritten

Once brass could be made to order rather than by luck, entire sectors changed their habits. Maritime builders lined ships with corrosion-resistant fixtures that standardized performance. Architects used brass hardware on exposed structures. Musical instrument makers tuned trumpets and horns with predictable alloys. Clockmakers filed gear teeth in brass that wore smoothly against steel pinions. Low-value coinage stayed legible longer, which kept market exchanges honest.

Even medicine and art changed. Zinc oxide became a reliable pigment for glazes and, later, for painters who wanted a white that resisted darkening. Apothecaries stocked zinc compounds for skin conditions. None of these uses announced themselves loudly. All traveled quietly, like the vapor that started the story.

The Military Edge You Do Not See

Gunpowder warfare brought new demands. Brass cartridge cases needed spring and ductility. Artillery required valves and gauges that held tolerance. Naval fleets needed instruments that stayed true in salt air. Each of these jobs rewarded consistent zinc control. The metal's role hid behind steel barrels and iron plates, yet when supply faltered, armies noticed.

Myths, Names, and Meanings

Legends occasionally gave zinc a theatrical aura. Some texts described furnace spirits or metals that breathed and devoured. The harsh white smoke from overheated zinc ores fed these images. In truth, the danger came from irritation when zinc oxide dust filled the air. Makers learned to read the plume and work upwind. They learned that the metal's value lay not in magic, but in the way it gave copper a new personality.

As distillation settled the question and metallic zinc appeared as a silver-gray mass, vocabulary tightened. Miners labeled ore. Refiners marked ingots. Brass founders posted ratios above the charge room. The element stepped out of the shadows and into the ledger.

The Element That Built a Bridge Between Ages

Look back across the arc. First came an accident in a hot room where copper met the right stone. A bright alloy appeared and won a place on temple doors and ship rails. Then came a long apprenticeship to vapor and seal, taught by furnaces in India and China and learned elsewhere over centuries. Distillation turned luck into schedule. Trade carried ingots and advice across mountains and harbors. Isotope science traced those paths like dew revealing a spider's web at dawn.

At each step, zinc changed other stories more than it claimed its own. It made coinage more durable, fleets more reliable, instruments more precise, and markets more trustworthy. It served quietly, and in serving, it shaped power.

The early metallurgists were right to watch the vapor with respect. They had found a partner that demanded patience and rewarded order. In an age that celebrated the metals you could see, touch, and weigh, zinc proved that transformative forces sometimes move in a thinner state. When captured with care, they become solid enough to hold empires together.

Chapter 14
The Phoenician Spark — Traders Who Spread Bronze Across the World

Waves slap wooden hulls in the bustling port of Tyre as ships bob impatiently, ready to carry the treasures of a small but mighty people across the Mediterranean. Bronze ingots shaped like ox hides thud against the decks as laborers haul them aboard. Seagulls circle overhead, shrieking for scraps in a chaos that smells of salt, resin, and hot metal. Near the harbor workshops, molten bronze cools in clay molds while merchants shout prices over clanking hammers. A scribe tallies cargo under the shade of a linen awning. Traders argue, sailors laugh, and the sea waits to distribute the work of Phoenician hands to distant shores.

Here, a spark is being struck, not through conquest or empire-building, but through trade and technology that will connect worlds.

Lighthouses Without Armies

The Phoenicians never ruled a massive land empire, nor did they dominate through overwhelming armies like Egypt or Assyria. Their power flowed from ships and metal. Located along the eastern Mediterranean coast, the cities of Tyre, Sidon, and Byblos thrummed with ingenuity, ambition, and maritime skill.

Where others built monuments to claim territory, the Phoenicians built networks. They carried copper, tin, and finished bronze goods from market to market. Their ships, swift and daring, threaded routes that linked Europe, Africa, and Asia into a commercial system.

Trade was their battlefield. Relationships were their treaties. Metal was their tool of influence.

Bronze for a Connected World

Bronze had already transformed ancient societies by the time Phoenician sailors hoisted their sails, but the alloy's great challenge remained access. Copper and tin rarely occurred together in nature. Civilizations that controlled one often lacked the other. The Phoenicians exploited this gap brilliantly.

They turned Cyprus into a metal heart of the Mediterranean. "Cyprus copper" became a brand and a lifeline. From island mines, the Phoenicians loaded heavy ingots onto cargo ships and delivered them to craftsmen across the region. In ports such as Ugarit and Crete, copper met tin, fires sparked, and bronze was born.

Their presence in the trade did more than deliver resources. It delivered knowledge. Bronze casting, alloy preferences, mold designs, and finishing techniques traveled alongside the merchants. Wherever Phoenician ships docked, workshops gained new ideas.

Carthage and the Forge of Influence

By the late first millennium BCE, the Phoenicians extended their reach far beyond the Levant. Their colonies—each a spark from the homeland—glowed along trading routes. The brightest flame was Carthage on the North African coast.

From Carthage, Phoenician traders sailed west into the Atlantic. They reached the mineral-rich coasts of Spain, including regions later associated with Tartessos. Here they found silver, lead, copper, and perhaps the prized tin essential for bronze. Often, they established mining operations or alliances that placed them at the center of resource exchange.

Carthage did not merely inherit Phoenician trade traditions. It scaled them up. Metallurgy became the root of Carthaginian wealth long before it became the cause of Punic War destruction.

Craft, Caravans, and Cultural Transfer

The Phoenicians understood that trade was never only about goods—techniques traveled too. Artifacts uncovered from Spain to Sardinia reveal bronze styles that blend local traditions with Levantine influence. Casting methods, decorative motifs, and mold shapes hint at a slow but powerful exchange that reshaped craft identity across continents.

Tools and trinkets became teachers. A single well-made hinge from Tyre could inspire new workshop standards in Italy. A shipment of Cypriot copper could help an Iberian foundry achieve more consistent alloys. Ideas traveled stowed inside cargo, unlocked by curious smiths far from the source.

Even alphabetic writing, standardized for commerce, moved across the Mediterranean like a new alloy—linking trade records and contracts across cultures.

The People Behind the Trade

Although often described collectively, the Phoenician world was driven by individuals—master navigators, merchants, and metal brokers whose names flicker in inscriptions.

King Hiram I of Tyre appears in multiple ancient sources as a leader who strengthened alliances built on timber and metal trade. Shipment by shipment, he expanded the economic reach of his city.

Inscriptions from Byblos reference royal sponsorship of foreign commerce. Clay tablets from Ugarit show Phoenician merchants negotiating with Mycenaean partners. Even Homeric epics mention Phoenician traders whose goods dazzled Greek heroes.

They were not anonymous profiteers. They were diplomats of the forge.

Archaeology Beneath the Waves

Much of what we know about Phoenician metallurgy lies underwater. Shipwrecks discovered near Uluburun, Cape Gelidonya, and other sites reveal cargoes stacked with copper ingots, tin pieces, bronze tools, and raw glass—often packed together as if intended for metallurgists eager to cast their next generation of artifacts.

The oxhide ingots themselves are clues to shared technology. Their shape—curved sides with handles at each corner—appears across the Mediterranean trade network. It suggests widespread standardization long before industrialization.

Each wreck tells a tale of trust, supply chains, and metallurgical intelligence.

A Quiet Revolution

Seaports may be loud, but their deepest impact often comes from the quiet cargo they move.

The Phoenicians changed warfare indirectly. By enabling others to obtain metal weapons efficiently, they empowered armies they never led. They changed economies by feeding marketplaces with standardized goods and stabilizing alloy quality. They changed art by supplying bronze to sculptors from Athens to Carthage.

Through their work, a scattered world became connected—a Mediterranean defined not by boundaries, but by routes.

When Conquest Overtook Commerce

Phoenician independence eventually eroded under the rise of larger empires. Assyrians, Babylonians, and Persians absorbed Phoenician cities. Later, Rome's rivalry with Carthage ended in fire and desolation.

Yet even conquerors relied on the networks the Phoenicians built. Roman bronze coins and fittings traveled pathways first charted by Tyrian sailors. Military forges in imperial camps shaped metal from regions discovered through Punic exploration.

Even in defeat, the Phoenicians remained the architects of Mediterranean commerce.

The Spark That Never Went Out

Archaeologists continue to recover bronze goods that testify to a world more connected than ancient chroniclers admitted. Each artifact holds a passport stamped by countless hands—miners, merchants, metalworkers, and sailors.

Stand again in the port of Tyre. Hear the hammers striking cooling bronze. Watch a merchant nod as a deal concludes. See sailors coil ropes and raise sails toward the horizon.

The Phoenicians never claimed to rule the world.

They simply supplied it. Their spark traveled not by sword, but by ship. And from that spark, civilizations lit their forges.

Chapter 15
The Silver That Saved Empires — Coins, Mints, and the Rise of Metal Money

The heat inside the royal mint feels almost alive. Bellows roar. Coals crackle. A worker grips a small, glowing disc of metal with iron tongs and places it delicately onto a carved die. Over it, another die is positioned with a symbol of power, a lion's head, a god's face, a king's pride. Then, with one swift strike of a heavy hammer, the image is stamped into history. The ringing of metal vibrates across the chamber as a new coin flies into a wooden coffer among dozens of others. In that single moment, raw silver becomes promise. It becomes trust. It becomes power that can move across mountains and seas without ever drawing a sword.

Money, in its simplest form, is belief. And in the ancient world, no force commanded belief more effectively than metal.

From Bartering to Standardization

For thousands of years, economies ran on barter, weighed ingots, or goods traded by negotiated value. But these systems slowed commerce. A goat and a sack of grain might be equal in one village yet disputed wildly in the next. As trade networks expanded and societies grew more complex, the need for a universal language of value became clear.

The breakthrough arrived in the kingdom of Lydia, located in western Anatolia during the seventh century BCE. Here, rulers discovered that a stable, regulated token of exchange made every transaction smoother. Lydia's earliest coins were struck from electrum, a natural alloy of gold and silver found in local riverbeds. Each coin bore a stamped emblem—often a lion's head—certifying authenticity and weight.

For the first time, value could be carried in a pocket, counted reliably, and recognized instantly even by strangers. Commerce accelerated. Markets widened. Money had been born not as an idea, but as a commodity in the form of a small metal disc.

Why Silver Became the Currency of Civilization

Gold dazzled with divine associations and royal prestige, but silver became the backbone of ancient economies. It was plentiful enough for broad circulation yet rare enough to maintain value. Most importantly, its physical properties met the demands of monetary exchange. Silver was durable. It resisted corrosion better than copper or iron. It could be divided and re-melted without losing essential worth.

The trust people placed in silver translated directly into political power. A ruler who minted reliable coinage could pay armies, build infrastructure, and collect taxes efficiently. Silver's steady glow sustained not only economies, but entire empires.

Greek Coins and the Spread of Monetary Innovation

As coinage spread from Lydia into the Greek world, each city-state adopted its own symbols. Athenian owls, Corinthian Pegasi, and Syracusan charioteers traveled across the Mediterranean as ambassadors of civic identity. These images evolved into widely recognized brands that guaranteed weight and purity. An Athenian tetradrachm could serve as currency from Spain to the Black Sea because merchants trusted its silver content.

Greek engineers refined minting methods, perfecting striking techniques to ensure crisp designs and consistent thickness. As alliances and rivalries sharpened, coinage helped fund both fleets and intellectual pursuits. Silver from Laurion financed Athens's ships during the Persian Wars, contributing to victory at Salamis. Philosophy discussed the nature of justice while silver paid for the walls that defended it.

Persia: The Mint of Administrators

When Persia built the largest empire the world had yet seen, it needed a system strong enough to unite dozens of languages and cultures. Darius I established a powerful minting standard: the gold daric and its silver companion, the siglos. These coins traveled the empire's vast network of roads and ports, from Egypt to India.

Persian taxation relied on standard payments rather than unpredictable tribute. Soldiers received steady salaries. Bureaucrats balanced ledgers. Monetary consistency made administration possible. The empire's wealth became fluid, mobile, and accountable in a way goods alone could never achieve. Where roads went, coins followed. Where coins went, Persia ruled.

Rome: Money as Empire

Rome elevated coinage into a tool of ideology. Emperors stamped their own faces on silver denarii, turning every transaction into a tiny act of loyalty. When a citizen bought bread, hired a mason, or paid taxes, the emperor's portrait witnessed the exchange.

At its height, Rome maintained hundreds of mints. Silver coins funded legions pressing into new territories and paid for the roads that carried spoils back to the capital. As long as Rome's silver remained pure, confidence in the empire remained unshaken. But trust was fragile.

The Metallurgy of Trust and the Temptation to Cheat

A silver coin is more than a promise—it is a test. Its weight and purity must match the standard behind it. Ancient rulers faced war costs and strained treasuries, and many succumbed to the temptation of debasement: reducing silver content while keeping the same face value.

At first, no one notices. The shine remains. The stamp appears unchanged. But markets always uncover truth. Money containing less silver commands less value. Prices rise. Inflation takes hold.

Rome offers the clearest caution. Over centuries, the silver content of the denarius dropped dramatically. Once nearly pure, by the third century CE it was mostly bronze with only a thin silver wash. Merchants resisted payment. Soldiers demanded higher wages. Economic chaos spread, weakening even Rome's military strength.

Metal manipulation—not only military defeat—helped undermine an empire built on confidence.

Counterfeiting: The Shadow Economy

Where money flows, deception follows. Ancient mints battled counterfeiters who plated base metal cores with silver or clipped edges from genuine coins. Scales and balances became as essential to trade as ships and wagons.

To combat fraud, rulers sharpened engravings, added intricate details, and enforced severe penalties.

Surviving coins reveal both official brilliance and illicit creativity.

Every forged coin reminds us that money commands faith only as long as its metal speaks truth.

How Technology Minted Power

Minting was itself a metallurgical achievement. Workers heated blanks to ideal temperatures so strikes filled dies cleanly. They experimented with alloys to endure repeated hammering. They refined molds to produce blanks of uniform mass.

Some cultures devised collar rings to prevent metal from spreading during striking. Others adopted rotating dies to speed production.

Innovation in coinage did more than improve currency. It reshaped economies. Coins enabled standardized wages, reliable tax collection, and accounting systems that supported expanding bureaucracies and trade networks.

Coins were the grease that kept vast imperial machines in motion.

Why Silver Mattered More Than Conquest

Silver did one thing no general could. It made strangers agree. In markets where no common language existed, a familiar coin created trust. Coins funded diplomacy, religion, infrastructure, and luxury. They recorded the ambitions of kings and the anxieties of crises. They measured the success of harvests and the failures of supply chains.

Consider the Silk Road. Caravans carried gold and silver not only as valuables, but as the tools of transaction. A silver coin connected two people separated by culture and distance. Metal money shortened the space between worlds.

When the Hammer Falls

Stand again inside the mint. Feel the pressure of a newly struck coin dropping warm into your palm. It carries weight both literal and symbolic. It contains the promise of bread, the loyalty of soldiers, and the value of a merchant's voyage. It is not merely silver—it is stability.

Across Lydia, Greece, Persia, and Rome, metal currency spared empires from the chaos of barter, forged trust across borders, and powered economies capable of sustaining cities and states larger than diplomacy alone could maintain.

The ring of a hammer striking a coin die echoes farther than the clang of swords. It declares order in a world that seeks it. Silver was civilization in solid form.

Empires stamped their ideals into metal because they understood this truth: The mightiest armies protect borders. The purest money protects futures.

Chapter 16
The Weapons of Heaven — Chinese Steel and the Crossbow Revolution

Fog clings to the morning fields like a shroud. War banners rise above the haze, painted with symbols of rival kingdoms locked in relentless struggle. An archer from the Qin army crouches behind a wooden barricade, his breath steady, his focus sharp. He slides a steel-tipped bolt into a compact bronze trigger mechanism mounted onto a stout wooden frame. With the slightest squeeze, the weapon snaps. The string releases. The bolt hisses through the mist and finds its mark. A distant soldier collapses before he even hears the sound that ended his life.

On this battlefield, the future of warfare has arrived. Metal is no longer just armor and blade. Now it is precision. Now it is distance. Now it is strategy. The crossbow, powered by bronze and sharpened by steel, changes the fate of nations long before empires rise from dust.

Across ancient China, metallurgy became more than craft. It became the mathematics of power.

Metal Forged for Conquest

China's mastery of iron and steel grew from necessity. By the Warring States period, roughly the fifth to third centuries BCE, the once-unified Zhou realm had fractured into competing kingdoms. States like Qin, Chu, Zhao, and Han poured resources into metallurgy to arm legions with better tools and weapons. Bronze, though revered and artistically unmatched, began to show its limits.

Iron smelting advanced rapidly. Craftsmen experimented with furnaces that reached higher temperatures. They learned to manipulate carbon levels by packing iron with charcoal and reheating it. Through trial and repetition, they discovered carburization, the process that transformed soft iron into hard steel.

Steel blades struck bronze shields and pierced them cleanly. Armor that once defended warriors now failed them. The kingdom that wielded steel first and best would rule the rest.

The Crossbow: A Mechanical Miracle

While swords and spears evolved through metallurgical advances, another innovation changed warfare more dramatically than any edge or point could. The crossbow fused ancient Chinese engineering, strategy, and metalworking into a single decisive instrument. Its roots stretch back to around the sixth century BCE, but by the height of the Warring States period, it had matured into a devastating battlefield equalizer.

Its power lay in a bronze trigger mechanism crafted with extraordinary precision. The mechanism held the bowstring under immense tension and released it with consistent force. Even untrained soldiers could draw, load, and fire with deadly accuracy. A farmer pressed into service might never learn a longbow, but he could master a crossbow within days.

Distance no longer belonged only to elite archers. Armor no longer shielded generals at a safe remove. Warfare became democratic—and far more deadly.

Standardization: The Science Behind Success

Archaeological discoveries at the mausoleum of China's first emperor, Qin Shi Huang, reveal how industrial thinking transformed military efficiency. Among the 8,000 warriors of the Terracotta Army, excavators found crossbow parts fashioned with remarkable uniformity. Trigger assemblies—nuts, levers, brackets—fit each other interchangeably.

Each Qin soldier carried a weapon built to identical measurements. If a mechanism jammed or snapped, soldiers replaced components on the field as easily as modern armies swap magazines.

Uniform production required structured workshops, skilled metallurgists, and tight state oversight. Metallurgy became the empire's quiet bureaucracy, operating behind walls of smoke and fire to ensure that every soldier held the future in his hands.

Terracotta Testimony

The weaponry buried beside the Terracotta Army does more than honor the troops. It offers a preserved lesson in craftsmanship. Many crossbow mechanisms remain sharp enough to function today. Analysis shows that their surfaces were polished and treated to resist corrosion. Their steel bolt tips were hardened through controlled carbon infusion. Their bronze triggers were engineered with exacting tolerances. Each detail underscores a simple truth: the Qin did not merely raise armies. They manufactured them.

A Race of Metallurgists

Metals shaped the politics of the age as much as diplomacy or alliances. Kingdoms competed fiercely to refine steel, sharpening their advantage with every furnace fired. The lines between workshop and battlefield blurred.

Zhao adopted steel armor that increased survivability without sacrificing speed. Han experimented with high-carbon weapon tips. Chu fortified borders with iron tools that built stronger defenses faster.

Yet Qin proved relentless. By combining standardized steel crossbows with disciplined infantry formations and efficient communication networks, it broke rival kingdoms one by one. The unification of

China under Qin Shi Huang was not only a political triumph—it was the victory of efficient metallurgy.

Weapon of the People, Fear of Kings

The crossbow introduced a terrifying new dimension to warfare: anonymity. A soldier no longer needed to face his enemy. Death traveled silently across the field. Elite warriors, once revered for close combat skill, could fall to an unseen foe crouching behind a barricade.

Rulers feared the crossbow because it empowered commoners. Some dynasties restricted ownership by law. The weapon could defend a village—or destabilize a throne. Control of metallurgy became control of power. Steel, more than any edict, enforced the hierarchy.

Hidden Knowledge: State Secrets of Steel

Chinese metallurgists guarded their techniques like sacred scripture. Crucible design, airflow control, quenching practices, and carbon management rarely appeared in surviving written form. Knowledge passed from master to apprentice, from state workshop to state workshop.

Iron produced in the north differed from that of the southern kingdoms, reflecting mineral sources and furnace adaptations. Some regions became famous for blades that retained an edge even against armor; others specialized in steel tools that carved terraces into mountainsides.

With ironworking spread across vast geography, innovation accelerated. The results radiated outward through Silk Road trade, touching the Middle East and Mediterranean centuries before Europe developed comparable steelmaking quality.

A Philosophy of War and Order

The crossbow's rise coincided with sweeping intellectual change. Legalist philosophers urged strict laws and discipline as the foundation for social order. Their ideas harmonized with Qin standardization: uniform weapons, uniform soldiers, uniform obedience.

Metallurgy did not merely support warfare. It shaped ideology. A society that could align thousands of craftsmen under one measurement could also align thousands of soldiers under one command. The steel-tipped bolt became the arrowhead of nation-building.

Steel After Qin

Although the Qin Empire endured only briefly, its metallurgical legacy persisted. Successive dynasties expanded steel production and diversified weaponry. The Han optimized crossbow deployment, issuing models capable of quick reloading and long-range precision.

Trade carried Chinese steel into neighboring regions. Even Rome, mighty though it was, prized iron shipments from the East. Later Chinese treatises described quenching techniques using brine, oil, or herbal mixtures—methods that anticipated global steel refinements centuries before Europe industrialized.

The crossbow endured as a central weapon in East Asia for more than 2,000 years, long after other regions replaced it with composite bows or early firearms.

The Legacy in the Museum and in Memory

Stand again on that foggy plain. See the Qin soldier's bronze trigger gleam beneath battle banners. Imagine him releasing a bolt that reshapes the balance of power across a continent.

Today, his weapon rests in museum cases, yet it still commands awe. Its engineering remains crisp. Its purpose remains clear. It tells the story of a world that advanced through metal—not by accident, but by relentless design.

Steel empowered armies and enforced empires. The crossbow democratized death and centralized control. Together, they proved that innovation in metal could shift the foundations of civilization.

Ancient China mastered the fire that shapes steel and the mechanism that directs it. Their weapons were born from heaven's elements. Their victories were written in steel.

Chapter 17
The Hidden Forges of Africa — Nok, Meroë, and Ingenuity Beyond the Desert

Drums echo beneath the dense canopy of sub-Saharan forest as night falls and stars pierce the sky. Flames roar inside clay furnaces tended by skilled blacksmiths whose hands remain steady even in the heat. Sparks rise like fireflies escaping the earth. Men and women gather to celebrate the birth of tools and weapons that will cut fields, shape homes, and defend the people who made them. Iron glows bright, then dims into strength. With every beat of the drum, the furnace breathes again, fed by charcoal and guided by knowledge older than written memory. Here, metal is not only technology. It is identity. It is power. It is life.

The Cradle of Iron Innovation

Africa's metallurgical story begins far from the deserts of the north and long before many historians once believed. The traditional narrative that metalworking spread into Africa from the Near East has steadily eroded as archaeological findings reveal deep indigenous innovation. In the heart of modern Nigeria, the Nok culture emerged around the last millennium BCE with a mastery that astonished twentieth-century archaeologists. Their iron production did not rely on borrowed concepts. It reflected technical independence built on resource knowledge, craftsmanship, and spiritual tradition.

Nok furnace designs varied—some stood above ground as cylinders, others were enclosed underground, each adapted to local environments. Charcoal fired the smelting process while natural drafts or crafted bellows supplied oxygen. The result was bloom iron, a spongy mass that required reheating and hammering on anvil stones to remove slag and strengthen the metal. The smiths' hammering did more than shape tools and weapons. It shaped communities.

Iron tools allowed the Nok to expand agriculture, increase yields, and feed larger settlements. Iron weapons protected trade routes and borders. Metallurgy became synonymous with advancement, and the Nok stood among the earliest pioneers of sub-Saharan technology.

Their artistry matched their metallurgy. Nok terracotta figures—stylized yet expressive, with intricate hairstyles, jewelry, and symbolic features—testify to cultural complexity supported by a stable economy. Technology created security. Art created identity.

Meroë, the City of Iron Kings

Far to the northeast, along a fertile bend of the Nile, another power rose in fire and ambition. The Kingdom of Kush, with its capital at Meroë, flourished after the decline of Egypt's New Kingdom. Between the sixth century BCE and the fourth century CE, Meroë transformed into a metallurgical powerhouse. While other African cultures smelted iron on smaller scales, Meroë industrialized it.

Archaeological surveys reveal vast furnace fields surrounding the ancient city. Mounds of slag stand like blackened hills—so immense that early travelers mistook them for volcanic rock. But these were the remains of centuries of production. Skilled engineers and blacksmiths harnessed the desert's resources with remarkable precision.

Meroitic furnaces used strong desert winds to reach temperatures required to reduce iron ore. Like Nok furnaces, they followed the bloomery model, producing iron that needed hammering to consolidate its strength. But the scale distinguished Meroë. They produced iron for farming tools, chariots, spearheads, nails, furniture fittings, and fine goods. Metallurgy fueled commerce, strengthened armies, and powered the kingdom's expansion along the Nile and into Africa's interior.

Unlike Egypt, which built permanence in stone, Meroë expressed power through adaptability—through metal that moved, worked, and shaped the future.

Trade Routes Through Wilderness and Empire

Africa was never isolated. Trade networks crossed the continent in every direction, linking metallurgical centers with the Mediterranean, Arabia, and distant kingdoms. Caravans crossed the Sahara long before written documentation captured their journeys. Salt, gold, ivory, and enslaved people were part of this commerce, but so were metal goods and metallurgical knowledge.

From Meroë, trade moved north into Egypt and west into the Nubian desert routes. Kushite rulers engaged diplomatically and militarily with Egyptian pharaohs, and for a time they ruled Egypt themselves as the 25th Dynasty. Their authority rested partly on iron that outmatched Egyptian bronze. When Rome conquered Egypt and advanced southward, they encountered Kushite archers and iron-armed warriors formidable enough to force diplomatic agreements. Treaties preserved Kushite autonomy long after other African states fell under Rome's influence.

Westward, Nok influence traveled into later Nigerian states, including Yoruba and Benin cultures, which developed exceptional bronze casting using lost-wax techniques. Southward, iron technologies shaped Bantu migrations that carried agriculture, language, and smelting knowledge across much of Africa.

Metal moved. Ideas moved. Africa stood at the crossroads of exchanges that reshaped world history.

The Blacksmith's Role as Maker and Mystic

Across many sub-Saharan cultures, blacksmiths held both authority and reverence. To master fire and metal was to engage with forces considered divine. The forge became a sacred space where earth, air, water, and fire combined to create objects alive with potential.

Ceremonies often accompanied smelting. Rituals guided when the furnace should be lit, how the ore should be placed, and who delivered the first hammer stroke. Music—drums, chants, and rhythmic song—blended with craft. Iron tools were not simply produced. They were born.

This fusion of technology and belief strengthened social hierarchies and protected metallurgical knowledge from misuse. In Africa, innovation thrived without separating scientific skill from spiritual worldview.

Colonial Shadows and Erased Achievements

For centuries, outsiders denied Africa's technological achievements. European scholars and explorers, guided by prejudice rather than evidence, insisted Africans could not have developed iron technology independently. They credited Egypt, the Middle East, or mythical invaders for every artifact uncovered south of the Sahara.

Nok terracotta sculptures were declared "too sophisticated" to be African. Slag fields at Meroë were dismissed as geological accidents because the idea of African industrial output clashed with colonial assumptions.

Yet evidence was immovable. Scientific studies—radiocarbon dating, metallurgical analysis, and isotopic testing—confirmed Africa's early and independent ironworking traditions. In some regions, iron was smelted centuries before similar technologies appeared in Europe. African innovation was not imitative. It was pioneering.

The tragedy lies not only in historical dismissal but in the disruption and loss of knowledge. Oral traditions fractured. Workshops closed. Skills eroded under foreign rule. Recognition is now growing, but history must continue to be corrected.

Fire That Built Societies

African metallurgy shaped not only economies and weapons but entire ways of life. Iron machetes cut forest to create farmland. Iron hoes turned soil that fed growing populations. Iron spearheads protected herds, villages, and borders. Terracotta figures preserved memory that stone never held.

From Nok to Meroë, metal created resilience and stability in lands of both abundance and hardship. Technology allowed people to adapt to their environments rather than yield to them. As communities expanded, trade grew, and cultures diversified. Every tool and weapon cast in clay furnaces fueled collective growth. The fire in the forge was the fire of the future.

Remembering the Hidden Forges

Today, museums display Nok terracottas under soft lights, though their true context was village life where smoke, song, and celebration surrounded metallurgical work. At Meroë, slender pyramids overlook vast fields of slag—silent evidence of innovation unmatched elsewhere in the ancient world.

Archaeologists continue to uncover furnace remains, tools, and trade artifacts that expand the story of Africa's metallurgical brilliance. Each discovery restores a piece of heritage long denied and reminds us that technological revolutions did not begin only along the Mediterranean or in Asian empires. They also began in the heart of Africa.

Stand again beneath the forest's shadow. Hear the drums marking the rhythm of the forge. See sparks rising into the night like tiny stars fleeing into the heavens.

Africa's ironworkers did not wait for knowledge to arrive from afar. They forged it themselves. Their flames shaped continents. Their craft reshaped history.

Chapter 18
The Masters of Tin – The Search for the Metal that Defined an Age

Rain slashes sideways across the jagged cliffs of Cornwall as miners swing picks into stubborn rock, searching for a vein of pale, promising metal hidden deep within the hills. Wind howls through tunnels carved by generations of men who knew that the fate of kingdoms rested on what they might unearth. In the dim glow of lantern light, the glint of ore sends hope rushing like blood through weary limbs. Tin. A metal that does not dazzle like gold, yet holds the power to transform soft copper into the mighty bronze that shapes empires. Every spark from the miners' tools is a reminder that survival, prosperity, and dominance can be struck from stone.

The Metal That Multiplied Power

Tin is a paradox of the ancient world. It lacks the brilliance and allure of gold, yet in the right hands becomes far more valuable. Mixed with copper at a careful ratio, tin creates bronze, a metal harder, sharper, and more durable than anything worked before. Bronze forms blades that bite rather than bend, tools that carve new paths through soil and stone, and armor that protects the ambitions of warriors.

Without tin, copper remains decorative. With tin, it becomes a force that builds civilizations.

As early as 3000 BCE, societies from the Mediterranean to Central Asia recognized this transformative pairing. The Bronze Age emerged as a revolution not only in weaponry but in agriculture, artistry, engineering, and trade. Yet a fundamental challenge shaped this new world: tin is scarce. It hides in veins scattered unpredictably across the earth. The communities that controlled it did not merely influence commerce. They influenced the trajectory of history.

Searching for the Sources of Bronze

Archaeologists and historians still debate the exact locations that supplied the Bronze Age with its lifeblood, but several regions stand out. Cornwall, battered by storms yet rich beneath the soil, is one of the most famous. Its mines would later feed European bronze industries for centuries, and some evidence suggests early exploitation may have begun even in

Bronze Age times. Across the sea in Iberia, tin deposits in Spain and Portugal were worked by local cultures whose networks reached far beyond their borders.

Western Europe, however, was only one part of the picture. Anatolia—modern Turkey—provided tin that moved toward the great empires of the Near East. In the east, deposits in the mountains of Central Asia supplied distant lands. Caravan networks—precursors to what would later become major Silk Road routes—braved deserts and peaks to move tin westward. The precious cargo traveled on donkey caravans for hundreds of miles, exchanged hand to hand through villages and palaces, its value rising with every step.

Tin became the glue that bound unexpected partners and the spark that ignited enduring conflict.

The Tin Road and the Exchange of Civilizations

Trade in tin created some of the earliest international supply chains. Merchant families in cities like Ugarit negotiated deals with suppliers from lands separated by mountains and seas. Ships carried ingots across the Mediterranean, stopping at islands and ports that became trade hubs long before written memory recorded their influence.

As bronze spread, so did cultural exchange. Techniques for alloying copper and tin were refined in workshops from Egypt to Mesopotamia. Craftsmen shared knowledge and learned to produce tools with consistent strength. Farmers plowed deeper. Builders raised monuments. Warriors stood firmer in battle. Civilization rose on bronze. And bronze rose on tin.

Power Defined by Ore

Tin's scarcity made it a political commodity. Rulers guarded mines with soldiers. When a kingdom gained access to a new deposit, its influence increased. When supply lines faltered, states staggered.

Hittite kings negotiated treaties that detailed metal movement as part of diplomacy. Egyptian records describe costly imports from distant lands they trusted—or sought to dominate. Mycenaean traders sailed farther than ever in pursuit of the silver-gray ore.

Access to tin meant access to dominance in every arena: battlefields, food production, monumental construction, and trade leverage. Bronze weapons and armor gave armies the advantage needed to conquer or resist. The metal supported economies by enabling more efficient labor. Even artistic expression flourished under bronze's durability. But scarcity brings risk.

Tracking Tin Through Time

Modern science has joined archaeology in unraveling the mysteries of tin's origins. Researchers analyze isotopic signatures in ancient artifacts to trace their geological birthplaces. A bronze axe discovered in Greece might tell a story that begins in the mountains of Afghanistan. A dagger buried in an Egyptian tomb may reveal a lineage that crosses the entire Mediterranean.

This forensic approach confirms what ancient artifacts imply. Tin never stayed still. It traveled through the hands of traders who bridged cultures separated by vast distances. Every object becomes a map, and every map reveals how interconnected the ancient world was—far earlier than once believed.

The Age That Needed Too Much

For centuries, trade routes held together the network that sustained the Bronze Age. Yet around 1200 BCE, the system faltered. Cities fell. Kingdoms disappeared. The Bronze Age Collapse remains a mystery, but tin almost certainly played a role.

As wars erupted and alliances fractured, trade routes became perilous. Tin supplies slowed or vanished. Workshops struggled to produce the metal goods that kept societies stable. Famine, migration, and economic breakdown spread.

When the ability to make bronze disappeared, so did the institutions that depended on it. The end of the Bronze Age did not arrive quietly. It arrived violently, as the world lost access to the resource that had built it.

From Crisis to Reinvention

Necessity drove innovation. As tin supplies dwindled, communities turned increasingly to iron. Iron ore was more common, though far more difficult to shape without advanced furnaces. Early iron tools were inferior to bronze, yet desperation forced adaptation. Over time, methods improved. The Iron Age dawned not through curiosity, but through survival.

Tin never lost its value, but its age of supremacy ended. The metal that once defined civilizations stepped aside for a more abundant rival.

The Legacy of Tin in the Story of Progress

The miners of Cornwall keep digging. The caravans of Central Asia keep crossing deserts. The ships in Iberian harbors keep loading ore under the wary eyes of traders who understand the cost of failure. Tin never sought glory, yet demanded respect, for without it copper remained humble.

Bronze was never simply a mixture of metals. It was a union of collaboration, ingenuity, and ambition. Tin taught societies how to work together through trade, and how they fell apart when supply collapsed.

The story of tin is the story of the Bronze Age itself: brilliant, fragile, and rooted in the earth. A small, inconspicuous element altered the fate of continents because it gave humanity the ability to shape its future. Through bronze, people forged tools that reshaped the land, weapons that shifted borders, and treasures that endured millennia.

The Masters of Tin were the masters of their age.

Empires rose on the promise of bronze. And bronze owed its power to a metal found in the hardest places, earned by the hardest labor, beneath unforgiving skies.

Chapter 19
The Lost Art of Tempering – Secrets Buried in Slag Heaps

The sun beats down on an open excavation site as archaeologists kneel in the dust, brushing aside loose soil from ancient refuse piles. These heaps are not filled with pottery or discarded household objects, but with the hardened remnants of fire: slag, the forgotten by-product of metal production. To untrained eyes, these lumps appear meaningless, the waste of a distant past. Yet when a trowel strikes something different—something glossy, metallic, almost glasslike—dust gives way to hardened steel, its structure warped by intense heat. In that moment, the silent slag heap speaks. It reveals that the people who once stood on this very ground possessed knowledge now obscured by time. Techniques once perfected in fire have vanished through secrecy, war, and the simple passing of ages.

These fragments hint at something extraordinary. Ancient craftspeople mastered complex heat treatments that transformed brittle iron into resilient steel. This was no accident. It was a science encoded in experience, built on generations of trial and error, and guarded in the memories of smiths who could gauge success by the heat's color alone.

Fire as a Tool of Transformation

Early metalsmiths learned that iron fresh from the furnace—bloom iron—was strong but inconsistent. Its structure was riddled with impurities. Striking it repeatedly while hot drove out slag, improved bonding, and refined the metal into a workable form. But what separated ordinary craftsmanship from mastery was the ability to manipulate iron's properties not only with the hammer but through precise adjustments in fire.

Heat treatment emerged as a hidden alchemy. By heating iron to specific temperatures and cooling it with intention, smiths changed its characteristics dramatically. A blade could become harder or softer, sharper or more flexible, depending on invisible transformations happening beneath its surface.

Ancient metallurgists lacked modern terminology, yet understood results. They watched color, listened to the sound of striking metal, and judged performance in battle or fieldwork. Knowledge lived in muscle memory and tradition. Every decision inside the forge shaped a weapon's fate.

Quenching: The Moment of the Blade's Soul

Perhaps the most dramatic of all heat treatments was quenching: heating a blade to a bright orange-yellow and plunging it into liquid. The effect was immediate and violent. Steam erupted, and metal that moments before had been soft hardened with breathtaking speed.

Water was the most common quenching medium, but cultures experimented endlessly. Brine delivered more predictable results. Oil slowed cooling and reduced breakage. Legends whispered of blades quenched in blood—not only superstition, but attempts to control temperature and chemistry in the absence of instruments.

Quenching hardened the surface, but left metal brittle if not treated further. A blade too hard could shatter like glass. Early blacksmiths recognized this fragility and developed a second vital step: tempering.

Tempering: Strength Through Controlled Weakness

Tempering meant reheating hardened steel to a lower temperature, watching colors shift through straw yellow, purple, and blue. Each hue offered a different balance of toughness and hardness. When done correctly, a blade gained resilience without losing strength. It would bend rather than snap, cut rather than crumble.

The extraordinary part is the precision this required. Smiths had no pyrometers or gauges. They read fire with their eyes and intuition. A sword might pass or fail based on a flicker of heat or a heartbeat too long in flame.

This knowledge marked the difference between weapons that saved lives and those that failed catastrophically.

Carburizing: Feeding Metal with Fire

Some early steel did not come from smelting high-carbon ore but through carburizing—forcing carbon into iron after smelting. Khmer and Chinese craftsmen packed iron bars in charcoal-filled clay boxes and heated them for hours or days. Carbon infiltrated the surface, transforming the iron into steel that could hold a sharper edge. Ancient India's wootz steel, used in iconic Damascus blades, relied on careful carbon control, producing microstructures admired by warriors and metallurgists alike.

Carburizing required patience and precision. Too little carbon left a blade soft. Too much made it brittle. The margin was narrow, but smiths who found it created materials that shaped empires.

Work Hardening: Strength at the Surface

Cold working—hammering metal after it cooled—increased hardness along the outer layer. This technique was especially useful for farm tools or armor that needed surfaces strong enough to resist wear while remaining flexible beneath. Ancient smiths wielded their hammers with intention, creating gradients of strength that matched each object's purpose.

Work hardening and tempering often complemented each other, forming complex sequences of heating, hammering, quenching, and reheating that worked in harmony.

A Global Competition of Secret Knowledge

Across the ancient world, mastery of heat treatment shaped military power. In China, carburized steel crossbow bolts pierced armor with relentless efficiency. In Japan, differential hardening produced katana blades with edges that resisted chipping while spines remained flexible. In Persia, steel-tipped spears and arrowheads gave cavalry an edge over bronze-equipped foes. In West Africa, bloomery smelters produced iron with remarkable carbon control that rivaled distant empires.

Techniques differed because environments, materials, and traditions differed. Innovation was everywhere, adapting with need and opportunity. What mattered most was that technology was never static. It evolved.

Silence in the Ruins

Heat treatment knowledge was often protected as a craft secret. Few had access to its methods. When wars destroyed cities or plague struck communities, skill could vanish in a single generation. What remained were fragments—broken weapons, cooled slag—and mysteries.

Slag heaps are now laboratories. Under microscopes, scientists examine metallic droplets to determine furnace temperatures. X-ray fluorescence reveals alloy composition. Microstructures etched into ancient steel expose clues of quenching and tempering cycles.

Experimental archaeologists rebuild furnaces and retrace ore sources. Their results occasionally match ancient brilliance—but often fall short. The secrets were too well guarded.

When Technology Outpaced Its Memory

Paradoxically, the Iron Age was not the apex of knowledge, but a beginning. Skills rose and eroded many times. Medieval Europe struggled to rediscover techniques mastered centuries earlier in Africa and Asia. Even Damascus steel, revered for its rippling patterns and legendary performance, vanished from production by the eighteenth century. Techniques abandoned became techniques unknown. Humanity often advances in waves—progress rising, then receding into shadow.

The Archaeologists' New Forge

Today's archaeometallurgists are detectives of fire. They analyze how different quenching liquids affect steel microstructures. They map isotopes to trace ore movement. They decode thermodynamics hidden in slag. Each breakthrough transforms dusty refuse into historical testimony.

One day, a misformed blade or overlooked furnace wall may reveal how a culture achieved brilliance beyond what we imagined. Slag heaps preserve failure and success side by side.

Lessons from a Lost Mastery

Stand again beside the archaeologists in heat and dust. Watch as they uncover a shard of steel, dull yet unbroken after centuries underground. See its surface glint under sunlight, evidence of a blade engineered to endure.

Ancient smiths wielded fire with courage and precision. They bent the earth's hardest material to their will, guided by instinct rather than instruments. Their achievements shaped survival, territorial expansion, and the tools that cultivated land and built civilizations. Heat treatment allowed humanity to control not only metal, but fate.

What remains buried in slag heaps is not merely waste. It is proof that innovation is fragile. It is a reminder that secrets disappear as quickly as they are discovered. Somewhere beneath soil and ruin, more answers wait silently. The forge is gone. The knowledge is gone. But the steel remains.

Chapter 20
Epilogue:
The Eternal Forge — How Lost Metals Continue to Shape the World

Molten metal pours from a towering furnace like a river of lightning, cascading into molds beneath a shower of sparks. Overhead cranes glide silently, moving glowing slabs of steel through the cavernous modern mill. The roar of vents, the rhythmic thunder of presses, and the hiss of cooling nozzles form a mechanical symphony. A metallurgist steps forward, visor reflecting orange fire, clipboard in hand, measuring carbon content and tensile strength with instruments ancestors could scarcely imagine. Yet the purpose is familiar. Transform metal into the backbone of civilization.

This moment, alive with flame and ingenuity, connects us to every forgotten forge and every vanished smith whose hands first learned to command the elements. Across time, the same heat that shaped the bronze axe, the Damascus sword, and the Persian spear now shapes the beams of skyscrapers and the chassis of electric vehicles. Humanity still stands before the fire, learning, adapting, daring. The story of metallurgy has not ended. It has only changed form.

Ancient Fires in a Modern World

Today's engineers wield lasers, electron microscopes, and computer simulations. They design alloys to resist corrosion beneath oceans, support bridges over rivers, or shield spacecraft from heat that rivals the temperature of the sun. Yet every discovery traces its lineage to breakthroughs born in distant centuries.

When a high-speed train glides along steel rails, it relies on carbon control perfected by forgotten cultures. When jet engines endure blistering friction, they employ alloys born from the same thinking that gave bronze its value. When surgical instruments hold a precise edge, they echo the experiments of smiths who tempered blades by instinct and firelight. Even the invisible—the steel skeleton inside a building or the aluminum frame of a plane—absorbs knowledge refined across millennia.

We stand on a mountain built from ore dug by ancient miners and shaped by vanished hands.

The Rediscovery of Legendary Metals

Some secrets proved resilient. Others faded, leaving behind only artifacts and awe. Modern researchers continue to chase mysteries that once seemed solved.

Damascus steel, with its mesmerizing ripples and uncanny sharpness, remains a puzzle. Its origins lie in India's wootz steel, whose high-carbon ingots contain microstructures now studied under powerful microscopes. Despite years of effort, the mechanical properties that allowed those swords to slice silk mid-air or resist shattering have not been fully recreated.

Roman flexible glass, if it existed as described, may have been an early form of metallic glass or a work-hardened marvel lost in the political fears of emperors.

Persian crucible steel, Japanese differential hardening, and West African bloomery mastery continue to inspire experiments in university labs and artisan forges.

The world keeps reaching backward to find paths forward. Ancient technologies still whisper possibilities.

Sustainability and the Future of the Forge

Modern metallurgy faces challenges earlier societies never confronted: finite resources, global demand, and environmental care. Every skyscraper beam, reinforced road, bolt, or bearing costs energy and material that must be sourced responsibly.

Engineers explore new frontiers: green steelmaking powered by hydrogen, recycled metals that rival raw ore, and alloys designed for longevity to reduce waste. Yet innovation remains rooted in lessons learned long ago. Ancient smelters understood airflow, fuel quality, ore purity, and furnace insulation. They knew that a breeze too strong or a coal too raw could ruin an entire batch.

Today's factories automate those refinements, but the principles persist. Metallurgy remains a dialogue between nature's constraints and human creativity.

Fire Creates, Fire Destroys

Every civilization that mastered metal discovered the same double-edged truth. The forge births both plow and spear, bridge and cannon. Ancient empires learned that technological power requires ethical clarity. Today's world faces that responsibility on a scale unimaginable to the smiths of Meroë or Harappa.

Steel forms surgical scalpels that save lives, but also missiles that take them. Precious metals power renewable technologies, yet mining them strains ecosystems. Progress demands vigilance—and wisdom.

The March from Bronze to Silicon

Though our age revolves around digital circuits and data streams, those technologies rely on metals more deeply than ever. Copper moves electricity through every wire. Nickel, cobalt, and manganese fill the batteries that drive new industries. Rare earth metals, difficult to refine, enable magnets in wind turbines and lenses in satellites.

We call it a silicon age, but silicon depends on aluminum housings, steel frameworks, copper connections. A smartphone holds the periodic table in miniature—a monument to mining, smelting, refining, and shaping. Every technological leap still begins with the same question ancient smiths asked: How can we make materials do more than they should?

Knowledge Lost, Knowledge Found

Experts who study ancient metals spend years reconstructing furnace designs, analyzing slag, and re-creating metalwork of long-lost cultures. Their findings rarely sit in isolation. Instead, they feed into contemporary innovation. Patterns in Damascus steel help improve advanced composites. Ancient bronze composition informs corrosion-resistant marine alloys. African bloom furnaces inspire more efficient rural smelting techniques.

History is a laboratory. Its failed experiments and forgotten successes hold keys to resilience and resourcefulness.

Fire, Water, Metal, and the Human Will

Even fields far from metallurgy feel its influence. Water engineering—once dependent on stone channels and wooden wheels—now relies on metal turbines, dams, and pipelines. Ancient hydraulic mastery, such as Roman aqueducts and Persian qanats, merges with modern metallurgy to solve twenty-first-century problems: drought, irrigation, clean water access.

The same elements that powered ancient empires now support global survival. Historic solutions often offer insight into future engineering because nature's challenges echo across time.

The Human Bond with the Forge

Stand in the modern mill again and focus not on machines, but on people: the metallurgist adjusting computer settings, the technician testing tensile strength, the crane operator moving glowing ingots with delicate precision. Their tools are new. Their struggles are ancient. Their pride is universal.

Blacksmiths from the Nok to the samurai would understand that creation through fire is creation of the self, metal demands discipline, metal rewards patience, and metal remembers effort.

The Eternal Pursuit

As long as humanity seeks progress, the forge will not go dark. Doctors will need scalpels that stay sharp. Astronauts will need hulls that endure impossible conditions. Builders will need beams that withstand earthquakes and storms. Farmers will need tools that cultivate land while conserving it.

The future is not post-metal. It is post-limitation. New alloys wait to be invented. New methods wait to be refined. New resources wait to be discovered or reused.

Yet the central lesson of this book remains clear. Innovation is fragile. When knowledge is lost, humanity spends centuries reclaiming it. We must protect not only what we learn, but how we learned it.

The Legacy of the Forge

In every age, a furnace has glowed somewhere in the world. Sparks have risen into the sky like prayers. The forge has always been a companion of civilization, a symbol of perseverance, curiosity, and the courage to transform the raw into the remarkable.

From stones that fell from heaven to furnaces that reshape mountains, we have chased mastery of fire not to conquer nature but to rise with it.

Ancient metals armed armies, built monuments, and electrified imagination. Their stories endure because humanity continues to evolve through the same hunger: to make what does not yet exist, to improve what already does, and to understand what once was lost.

The flames that once lit desert workshops, mountain foundries, and forest smelters now illuminate laboratories and industrial skylines. Yet at their heart burns the same spark that drove the first smith to strike iron against stone. Progress remains a fire we must feed with care. Wisdom remains the hammer that shapes it. And as long as metal flows like molten sunlight from the furnace, the ancient forge continues: eternal, transformative, and profoundly human.

Made in the USA
Monee, IL
04 November 2026

67384794R00089